Linear Transformation

Mathematical Engineering, Manufacturing, and Management Sciences

Series Editor
Mangey Ram
Professor, Assistant Dean (International Affairs), Department of Mathematics, Graphic Era University, Dehradun, India

The aim of this new book series is to publish the research studies and papers that bring up the latest development and research applied to mathematics and its applications in the manufacturing and management sciences areas. Mathematical tools and techniques are the strength of engineering sciences. They form the common foundation of all novel disciplines as engineering evolves and develops. This series will include a comprehensive range of applied mathematics and its application in engineering areas such as optimization techniques, mathematical modelling and simulation, stochastic processes and systems engineering, safety-critical system performance, system safety, system security, high assurance software architecture and design, mathematical modelling in environmental safety sciences, finite element methods, differential equations, and reliability engineering.

Recent Advancements in Graph Theory
Edited by N. P. Shrimali and Nita H. Shah

Mathematical Modeling and Computation of Real-Time Problems:
An Interdisciplinary Approach
Edited by Rakhee Kulshrestha, Chandra Shekhar, Madhu Jain, and Srinivas R. Chakravarthy

Circular Economy for the Management of Operations
Edited by Anil Kumar, Jose Arturo Garza-Reyes, and Syed Abdul Rehman Khan

Partial Differential Equations: An Introduction
Nita H. Shah and Mrudul Y. Jani

Linear Transformation: Examples and Solutions
Nita H. Shah and Urmila B. Chaudhari

Matrix and Determinant: Fundamentals and Applications
Nita H. Shah and Foram A. Thakkar

Non-Linear Programming: A Basic Introduction
Nita H. Shah and Poonam Prakash Mishra

For more information about this series, please visit: https://www.routledge.com/ Mathematical-Engineering-Manufacturing-and-Management-Sciences/book-series/ CRCMEMMS

Linear Transformation
Examples and Solutions

Nita H. Shah and Urmila B. Chaudhari

CRC Press
Taylor & Francis Group
Boca Raton London New York

CRC Press is an imprint of the
Taylor & Francis Group, an **informa** business

First edition published 2021
by CRC Press

6000 Broken Sound Parkway NW, Suite 300, Boca Raton, FL 33487-2742

and by CRC Press

2 Park Square, Milton Park, Abingdon, Oxon, OX14 4RN

© 2021 Nita H. Shah and Urmila B. Chaudhari

CRC Press is an imprint of Taylor & Francis Group, LLC

Library of Congress Cataloging-in-Publication Data
Names: Shah, Nita H., author. | Chaudhari, Urmila B., author.
Title: Linear transformation : examples and solutions / Nita H. Shah and Urmila B. Chaudhari.
Description: First edition. | Boca Raton : CRC Press, 2021. |
Series: Mathematical engineering, manufacturing, and management sciences |
Includes bibliographical references and index.
Identifiers: LCCN 2020040003 (print) | LCCN 2020040004 (ebook) |
ISBN 9780367613259 (hardback) | ISBN 9781003105206 (ebook)
Subjects: LCSH: Algebras, Linear—Problems, exercises, etc. | Linear operators.
Classification: LCC QA184.5 .S53 2021 (print) | LCC QA184.5 (ebook) |
DDC 512/.5—dc23
LC record available at https://lccn.loc.gov/2020040003
LC ebook record available at https://lccn.loc.gov/2020040004

ISBN: 978-0-367-61325-9 (hbk)
ISBN: 978-1-003-10520-6 (ebk)

Typeset in Times
by codeMantra

Contents

Preface

The vital role of linear transformations in linear algebra is precisely imitated by the fact that it is the emphasis of all linear algebra books. Research on teaching and learning linear algebra has been studied by several research groups in the last 10 years. Researchers concur that in spite of its many applications, this is a hard subject for students, and many of the difficulties that students face have been emphasized and explained in terms of different theoretical frameworks. In the subject of linear algebra, linear transformations have received a lot of attention by researchers because of their importance in applications and the issues students face when learning them. It has been observed that students struggle when asked to find a linear transformation in a geometric context beginning with images of basis vectors; they have problems using systemic reasoning and using visualization to determine the transformations; students show a tendency to use intuitive models when working geometrically and conceptualizing transformations as functions. Keeping all these points in mind, we turn our interest to the area of linear transformations.

MATLAB® is a registered trademark of The MathWorks, Inc. For product information, please contact:

The MathWorks, Inc.
3 Apple Hill Drive
Natick, MA 01760-2098 USA
Tel: 508-647-7000
Fax: 508-647-7001
E-mail: info@mathworks.com
Web: www.mathworks.com

Acknowledgements

First, at this stage, I would like to extend my sincere gratitude and thank my PhD guide Prof. (Dr.) Nita H. Shah for her constant encouragement and support which I cannot describe in words. She is an inspiration for me as I have witnessed her great multidisciplinary knowledge and enthusiasm. From her, I always learnt to be dedicated, energetic, punctual, sharp, and patient.

I am grateful to my all siblings and parents Bipinchandra and Savitaben, who have provided me with thorough moral and emotional support in my life. I am also grateful to my other family members and friends who have supported me along the way.

I express my sincerest gratitude to my husband Dr. Mrudul for his positive suggestions to improve the standard of this book. Your unbeatable support has made my journey of writing this book a satisfactory success, which I will always cherish. Also, I am very thankful to my mother-in-law Purnaben and father-in-law Yogeshkumar for their constant support.

Dr. Urmilaben B. Chaudhari

Authors

Nita H. Shah, PhD, is HOD of the Department of Mathematics in Gujarat University, India. She received her PhD in statistics from Gujarat University in 1994. She is a post-doctoral visiting research fellow of University of New Brunswick, Canada. Prof. Shah's research interests include inventory modelling in supply chains, robotic modelling, mathematical modelling of infectious diseases, image processing, dynamical systems and their applications, and so on. Prof. Shah has published 13 monographs, 5 textbooks, and 475+ peer-reviewed research papers. Four edited books were published by IGI Global and Springer with coeditor Dr. Mandeep Mittal. Her papers are published in high-impact Elsevier, Inderscience, and Taylor & Francis journals. She is an author of 14 books. On Google Scholar, she has more than 3070 citations and the maximum number of citations for a single paper is over 174. The H-index is 24 as of March 2020 and i-10 index is 77. She has guided 28 PhD students and 15 MPhil students. Seven students are pursuing research for their PhD. She has travelled to United States, Singapore, Canada, South Africa, Malaysia, and Indonesia to give talks. She is vice-president of the Operational Research Society of India. She is council member of Indian Mathematical Society.

Urmila B. Chaudhari, PhD, received her PhD in 2018 in mathematics. She has 8 years of teaching experience at Engineering Degree College, she is a lecturer in the Government Polytechnic, Dahod, Gujarat, India. Her research interests are in the fields of supply chain inventory modelling for different payment options. She has published 21 articles in international journals, including Taylor & Francis, Springer, IGI Global, Growing Sciences, Revista Investigacion Operacional, Inderscience, International Knowledge Press, IJOQM, and AMSE. She has also published five book chapters in international edited books. She is the author of one international book.

1 Linear Transformations of Euclidean Vector Space

In this chapter, we will start with the study about the function of the form $w = f(X)$, where X is the independent vector in R^n and w is a dependent variable in R^m. We will concentrate on a special case as the transformation from R^n to R^m. A linear transformation is fundamental in the study of linear algebra. Hence, we will study linear transformation and different types of operators such as reflection operator, orthogonal projection operator, rotation operator, dilation operator, contraction operator, and shear operator. Many applications are related to the operator in the field of physics, engineering, computer graphics, social sciences, and various branches of mathematics. We will introduce the composition of two or more linear transformations in this chapter. With the help of this composition, one can analyse the vector in the space after applying the different types of linear transformations. We will start with the concept of transformation then afterwards the content.

1.1 LINEAR TRANSFORMATIONS FROM R^n TO R^m

1.1.1 TRANSFORMATIONS FROM R^n TO R^m

If the domain of the function f is R^n and co-domain of the function f is R^m, then f is called transformation from R^n to R^m. It can be denoted by $f : R^n \rightarrow R^m$. In this case, when $n = m$, the transformation f is called an operator on R^n.

To demonstrate the transformation from R^n to R^m, suppose $f_1, f_2, f_3, f_4, \ldots, f_n$ are the real-valued functions of n real variables, say

$$\left.\begin{aligned}
\omega_1 &= f_1(x_1, x_2, x_3, \ldots, x_n) \\
\omega_2 &= f_2(x_1, x_2, x_3, \ldots, x_n) \\
\omega_3 &= f_3(x_1, x_2, x_3, \ldots, x_n) \\
& \quad . \quad \quad . \\
& \quad . \quad \quad . \\
\omega_m &= f_m(x_1, x_2, x_3, \ldots, x_n)
\end{aligned}\right\}. \tag{1.1}$$

If this transformation is denoted by T, then $\mathrm{T} : R^n \rightarrow R^m$ and $T(x_1, x_2, x_3, \ldots, x_n) = (\omega_1, \omega_2, \ldots, \omega_m)$.

Example 1.1

If the transformation $T: R^3 \rightarrow R^4$ is defined by the equations $\omega_1 = x_1$, $\omega_2 = x_2 - x_3$, $\omega_3 = 3x_1x_3$, and $\omega_4 = x_1^2 + x_2^3$, then the image point of (x_1, x_2, x_3) is expressed as a transformation $T(x_1, x_2, x_3) = (x_1, x_2 - x_3, 3x_1x_3, x_1^2 + x_2^3)$.

Thus, for an instant, the image of $(1,0,1)$ is $T(1,0,1) = (1, -1, 3, 1)$.

1.1.2 LINEAR TRANSFORMATIONS FROM R^n TO R^m

In equation (1.1), if all equations are linear, then the transformation $T: R^n \rightarrow R^m$ is called a linear transformation. In this case, when $n = m$, the transformation T is called a linear operator on R^n. Therefore, the linear transformation $T: R^n \rightarrow R^m$ is defined by the equations

$$\left.\begin{array}{l} \omega_1 = a_{11}x_1 + a_{12}x_2 + a_{13}x_3 + \cdots + a_{1n}x_n \\[2mm] \omega_2 = a_{21}x_1 + a_{22}x_2 + a_{23}x_3 + \cdots + a_{2n}x_n \\[2mm] \qquad . \qquad . \\[1mm] \qquad . \qquad . \\[2mm] \omega_m = a_{m1}x_1 + a_{m2}x_2 + a_{m3}x_3 + \cdots + a_{mn}x_n \end{array}\right\}. \qquad (1.2)$$

If this linear transformation is denoted by T, then $T: R^n \rightarrow R^m$ and $T(x_1, x_2, x_3, \ldots, x_n) = (\omega_1, \omega_2, \ldots, \omega_m)$.

Example 1.2

If the transformation $T: R^2 \rightarrow R^3$ is defined by the equations $\omega_1 = x_1$, $\omega_2 = x_2 - x_1$, and $\omega_3 = 3x_1 + x_2$, then the image point of (x_1, x_2) is expressed as a linear transformation $T(x_1, x_2) = (x_1, \ x_2 - x_1, \ 3x_1 + x_2)$. For example, the image point of $(1, -1)$ is $T(1, -1) = (1, -2, 2)$.

1.2 MATRIX REPRESENTATION OF LINEAR TRANSFORMATIONS

Matrix representation of equation (1.2) is

$$\begin{bmatrix} \omega_1 \\ \omega_2 \\ . \\ . \\ . \\ \omega_m \end{bmatrix} = \begin{bmatrix} a_{11} & a_{12} & a_{13} \ldots a_{1n} \\ a_{21} & a_{22} & a_{23} \ldots a_{2n} \\ . \\ . \\ . \\ a_{m1} & a_{m2} & a_{m3} \ldots a_{mn} \end{bmatrix} \begin{bmatrix} x_1 \\ x_2 \\ . \\ . \\ . \\ x_n \end{bmatrix} \qquad (1.3)$$

or more briefly by

$$W = AX. \tag{1.4}$$

In equation (1.4), W is a vector in R^m, X is a vector in R^n, the matrix $A = \{a_{ij}\}$ is called the **standard matrix** for the linear transformation T, and T is called multiplication by A.

Example 1.3

Find the standard matrix for the linear transformation defined by the equations $\omega_1 = 2x_1 - 3x_2 + x_4$ and $\omega_2 = 3x_1 + 5x_2 - x_3$.

Solution

Let linear transformation of the above equation be denoted by $T: R^4 \to R^2$,
$$T(x_1, x_2, x_3, x_4) = (w_1, w_2) = (2x_1 - 3x_2 + x_4, 3x_1 + 5x_2 - x_3)$$
Therefore, the standard matrix T is

$$T = \begin{bmatrix} 2 & -3 & 0 & 1 \\ 3 & 5 & -1 & 0 \end{bmatrix}.$$

Example 1.4

Find the standard matrix for the linear transformation T defined by the formula
$$T(x_1, x_2) = (x_2, -x_1, x_1 + 3x_2, x_1 - x_2).$$

Solution

The standard matrix for the linear transformation $T(x_1, x_2) = (x_2, -x_1, x_1 + 3x_2, x_1 - x_2)$ is

$$T = \begin{bmatrix} 0 & 1 \\ -1 & 0 \\ 1 & 3 \\ 1 & -1 \end{bmatrix}.$$

Example 1.5

Find the standard matrix for the linear operator T defined by the formula
$$T(x_1, x_2, x_3) = (4x_1, 7x_2, -8x_3).$$

Solution

The standard matrix for the linear operator $T(x_1, x_2, x_3) = (4x_1, 7x_2, -8x_3)$ is

$$T = \begin{bmatrix} 4 & 0 & 0 \\ 0 & 7 & 0 \\ 0 & 0 & -8 \end{bmatrix}.$$

Definition: Zero Transformations from R^n to R^m

If 0 is the $m \times n$ zero matrix and $\overline{0}$ is the zero vector in R^n, then for every vector x in R^n, $T_0(x) = 0 \cdot x = \overline{0}$. It is called a zero transformation from R^n to R^m.

Definition: Identity Operator on R^n

If I is the $n \times n$ identity matrix, then for every vector x in R^n, $T_I(x) = I \cdot x = x$. It is called an identity operator on R^n. Sometimes the identity operator is denoted by I.

1.3 DIFFERENT TYPES OF OPERATORS AND THEIR COMPOSITION

Most important linear operators on R^2 and R^3 are those that produce reflections, projections, rotation, dilation, contraction, and shears. Now, such types of operators in R^2 and R^3 can be discussed as follows.

1.3.1 REFLECTION OPERATOR

Consider the operator $T : R^2 \rightarrow R^2$ that maps each vector into its symmetric image about the different axes is shown in Table 1.1 (Figures 1.1–1.3).

Consider the operator $T : R^3 \rightarrow R^3$ that maps each vector into its symmetric image about the different planes is shown in Table 1.2.

In general, operators on R^2 and R^3 that map each vector into its symmetric image about some line or plane are called **reflection operators**.

1.3.2 ORTHOGONAL PROJECTION OPERATOR

Consider the operator $T : R^2 \rightarrow R^2$ that maps each vector into its orthogonal projection about the x-axis and y-axis is shown in Table 1.3 (Figures 1.4 and 1.5).

Consider the operator $T : R^3 \rightarrow R^3$ that maps each vector into its orthogonal projection about the different planes is shown in Table 1.4.

In general, an orthogonal projection operator on R^2 or R^3 is any operator that maps each vector into its orthogonal projection on a line or plane through the origin.

1.3.3 ROTATION OPERATOR

An operator that rotates the vector in R^2 through an angle θ is called a rotation operator. If the point (x, y) is rotated by an angle θ, then the point reaches at the angle θ from the positive x-axis. So, from the basic concept of trigonometry, the polar form of the (x, y) point is $(r\cos\phi, r\sin\phi)$. After rotation through an angle θ, it will reach at $(r\cos(\theta + \phi), r\sin(\theta + \phi))$. If T is the rotation operator, then $T(x, y) = (r\cos(\theta + \phi), r\sin(\theta + \phi))$.

Using trigonometric identities on $\omega_1 = r\cos(\phi + \theta)$, $\omega_2 = r\sin(\phi + \theta)$ yields,

$$\omega_1 = r\cos\phi\cos\theta - r\sin\phi\sin\theta, \quad \omega_2 = r\sin\theta\cos\phi + r\sin\phi\cos\theta. \qquad (1.5)$$

TABLE 1.1

Reflection Operator on R^2

Operator	Illustration	Equation	Transformation and Its Standard Matrix
Reflection about the x-axis on R^2		$\omega_1 = x,$ $\omega_2 = -y$	$T(x,y) = (x,-y),$ $T = \begin{bmatrix} 1 & 0 \\ 0 & -1 \end{bmatrix}$

FIGURE 1.1 Reflection about the x-axis.

| Reflection about the y-axis on R^2 | | $\omega_1 = -x,$ $\omega_2 = y$ | $T(x,y) = (-x,y),$ $T = \begin{bmatrix} -1 & 0 \\ 0 & 1 \end{bmatrix}$ |

FIGURE 1.2 Reflection about the y-axis.

| Reflection about the line $y = x$ on R^2 | | $\omega_1 = y,$ $\omega_2 = x$ | $T(x,y) = (y,x),$ $T = \begin{bmatrix} 0 & 1 \\ 1 & 0 \end{bmatrix}$ |

FIGURE 1.3 Reflection about the line $y = x$.

Now, substituting $x = r\cos\phi$ and $y = r\sin\phi$ in equation (1.5), we get $\omega_1 = x\cos\theta - y\sin\theta$ and $\omega_2 = x\sin\theta + y\cos\theta$. These equations are linear; therefore, T is a linear operator (Figure 1.6; Tables 1.5 and 1.6).

1.3.4 CONTRACTION AND DILATION OPERATORS

An operator on R^2 or R^3 that compresses each vector homogeneously towards the origin from all directions is called a contraction operator, whereas an operator on

TABLE 1.2
Reflection Operator on R^3

Operator	Equation	Transformation and Its Standard Matrix
Reflection about the xy-plane on R^3	$\omega_1 = x, \omega_2 = y, \omega_3 = -z$	$T(x,y,z)=(x,y,-z), T = \begin{bmatrix} 1 & 0 & 0 \\ 0 & 1 & 0 \\ 0 & 0 & -1 \end{bmatrix}$
Reflection about the xz-plane on R^3	$\omega_1 = x, \omega_2 = -y, \omega_3 = z$	$T(x,y,z)=(x,-y,z), T = \begin{bmatrix} 1 & 0 & 0 \\ 0 & -1 & 0 \\ 0 & 0 & 1 \end{bmatrix}$
Reflection about the yz-plane on R^3	$\omega_1 = -x, \omega_2 = -y, \omega_3 = z$	$T(x,y,z)=(-x,y,z), T = \begin{bmatrix} -1 & 0 & 0 \\ 0 & 1 & 0 \\ 0 & 0 & 1 \end{bmatrix}$

TABLE 1.3
Orthogonal Projection on R^2

Operator	Illustration	Equation	Linear Transformation and Its Standard Matrix
Orthogonal projection on the x-axis in R^2		$\omega_1 = x, \omega_2 = 0$	$T(x,y)=(x,0), T = \begin{bmatrix} 1 & 0 \\ 0 & 0 \end{bmatrix}$
Orthogonal projection on the y-axis in R^2		$\omega_1 = 0, \omega_2 = y$	$T(x,y)=(0,y), T = \begin{bmatrix} 0 & 0 \\ 0 & 1 \end{bmatrix}$

FIGURE 1.4 Orthogonal projection on the x-axis.

FIGURE 1.5 Orthogonal projection on the y-axis.

TABLE 1.4
Orthogonal Projection on R^3

Operator	Equation	Linear Transformation and Its Standard Matrix
Orthogonal projection on the xy-plane in R^3	$\omega_1 = x, \omega_2 = y, \omega_3 = 0$	$T(x,y,z)=(x,y,0), T = \begin{bmatrix} 1 & 0 & 0 \\ 0 & 1 & 0 \\ 0 & 0 & 0 \end{bmatrix}$
Orthogonal projection on the xz-plane in R^3	$\omega_1 = x, \omega_2 = 0, \omega_3 = z$	$T(x,y,z)=(x,0,z), T = \begin{bmatrix} 1 & 0 & 0 \\ 0 & 0 & 0 \\ 0 & 0 & 1 \end{bmatrix}$
Orthogonal projection on the yz-plane in R^3	$\omega_1 = 0, \omega_2 = y, \omega_3 = z$	$T(x,y,z)=(0,y,z), T = \begin{bmatrix} 0 & 0 & 0 \\ 0 & 1 & 0 \\ 0 & 0 & 1 \end{bmatrix}$

TABLE 1.5
Rotation Operator on R^2

Operator	Illustration	Equation	Linear Transformation and Its Standard Matrix
Rotation through an angle θ		$\omega_1 = x\cos\theta - y\sin\theta,$ $\omega_2 = x\sin\theta + y\cos\theta$	$T(x,y)=\begin{pmatrix} x\cos\theta - y\sin\theta, \\ x\sin\theta + y\cos\theta \end{pmatrix},$ $T = \begin{bmatrix} \cos\theta & -\sin\theta \\ \sin\theta & \cos\theta \end{bmatrix}$

FIGURE 1.6 Rotation through an angle θ.

R^2 or R^3 that stretches vector homogeneously towards the origin from all directions is called a dilation operator. These are shown in Table 1.7 (Figures 1.7 and 1.8) and Table 1.8.

1.3.5 SHEAR OPERATOR

An operator on R^2 that moves each point parallel to the x-axis by the amount ky is called a shear in the x-direction. Similarly, an operator on R^2 that moves each point parallel to the y-axis by the amount kx is called a shear in the y-direction. These are shown in Table 1.9.

TABLE 1.6
Rotation Operator on R^3

Operator	Equation	Linear Transformation and Its Standard Matrix
Counterclockwise rotation about the positive x-axis through an angle θ	$\omega_1 = x,\ \omega_2 = y\cos\theta - z\sin\theta,$ $\omega_3 = y\sin\theta + z\cos\theta$	$T(x,y,z) = \begin{pmatrix} x, \\ y\cos\theta - z\sin\theta, \\ y\sin\theta + z\cos\theta \end{pmatrix},$ $T = \begin{bmatrix} 1 & 0 & 0 \\ 0 & \cos\theta & -\sin\theta \\ 0 & \sin\theta & \cos\theta \end{bmatrix}$
Counterclockwise rotation about the positive y-axis through an angle θ	$\omega_1 = x\cos\theta + z\sin\theta,\ \omega_2 = y,$ $\omega_3 = -x\sin\theta + z\cos\theta$	$T(x,y,z) = \begin{pmatrix} x\cos\theta + z\sin\theta, \\ y, \\ -x\sin\theta + z\cos\theta \end{pmatrix},$ $T = \begin{bmatrix} \cos\theta & 0 & \sin\theta \\ 0 & 1 & 0 \\ -\sin\theta & 0 & \cos\theta \end{bmatrix}$
Counterclockwise rotation about the positive z-axis through an angle θ	$\omega_1 = x\cos\theta - y\sin\theta,$ $\omega_2 = x\sin\theta + y\cos\theta,\ \omega_3 = z$	$T(x,y,z) = \begin{pmatrix} x\cos\theta - y\sin\theta, \\ x\sin\theta + y\cos\theta, \\ z \end{pmatrix},$ $T = \begin{bmatrix} \cos\theta & -\sin\theta & 0 \\ \sin\theta & \cos\theta & 0 \\ 0 & 0 & 1 \end{bmatrix}$

1.4   COMPOSITION OF TWO OR MORE TRANSFORMATIONS

If $T_1 : R^n \rightarrow R^k$ and $T_2 : R^k \rightarrow R^m$ are two linear transformations, then first apply T_1 a transformation on a vector that is in R^n and after that apply the second transformation T_2, i.e., $T_2 \circ T_1(X)$. The application of T_1 followed by T_2 produces a transformation from R^n to R^k. This transformation is called the composition of T_2 with T_1 and is denoted by $T_2 \circ T_1$ and $T_2 \circ T_1(X) = T_2(T_1(X))$, where $X \in R^n$.

Remark: Multiplying matrices is comparable to composing the corresponding linear transformation in the right-to-left order of the factors. If $T_1 : R^n \rightarrow R^k$ and $T_2 : R^k \rightarrow R^m$ are two linear transformations, then because the standard matrix for the composition $T_2 \circ T_1$ is the product of the standard matrices of T_2 and T_1, we have $[T_2 \circ T_1] = [T_2] \cdot [T_1]$.

TABLE 1.7
Contraction and Dilation Operators on R^2

Operator	Illustration	Equation	Linear Transformation and Its Standard Matrix
Contraction with factor k on R^2 $(0 \le k \le 1)$		$\omega_1 = kx,$ $\omega_2 = ky$	$T(x,y)=(kx,ky),$ $T = \begin{bmatrix} k & 0 \\ 0 & k \end{bmatrix}$

FIGURE 1.7　Contraction with factor k.

| Dilation with factor k on R^2 $(k \ge 1)$ | | $\omega_1 = kx,$ $\omega_2 = ky$ | $T(x,y)=(kx,ky),$ $T = \begin{bmatrix} k & 0 \\ 0 & k \end{bmatrix}$ |

FIGURE 1.8　Dilation with factor k.

TABLE 1.8
Contraction and Dilation Operators on R^3

Operator	Equation	Linear Transformation and Its Standard Matrix
Contraction with factor k on R^3 $(0 \le k \le 1)$	$\omega_1 = kx, \omega_2 = ky,$ $\omega_3 = kz$	$T(x,y,z)=(kx,ky,kz), T = \begin{bmatrix} k & 0 & 0 \\ 0 & k & 0 \\ 0 & 0 & k \end{bmatrix}$
Dilation with factor k on R^3 $(k \ge 1)$	$\omega_1 = kx, \omega_2 = ky,$ $\omega_3 = kz$	$T(x,y,z)=(kx,ky,kz), T = \begin{bmatrix} k & 0 & 0 \\ 0 & k & 0 \\ 0 & 0 & k \end{bmatrix}$

TABLE 1.9
Shear Operator

Operator	Equation	Linear Transformation and Its Standard Matrix
Shear in the x-direction on R^2	$\omega_1 = x + ky, \omega_2 = y$	$T(x,y) = (x + ky, y), T = \begin{bmatrix} 1 & k \\ 0 & 1 \end{bmatrix}$
Shear in the y-direction on R^2	$\omega_1 = x, \omega_2 = kx + y$	$T(x,y) = (x, kx + y), T = \begin{bmatrix} 1 & 0 \\ k & 1 \end{bmatrix}$

Example 1.6

Let $T_1 : R^2 \rightarrow R^2$ and $T_2 : R^2 \rightarrow R^3$ be the transformations given by $T_1(x,y) = (x+y, y)$ and $T_2(x,y) = (2x, y, x+y)$. Find the formula for $T_2 \circ T_1(x,y)$.

Solution

Let the standard matrix for $T_1(x,y) = (x+y, y)$ transformation is $T_1 = \begin{bmatrix} 1 & 1 \\ 0 & 1 \end{bmatrix}$ and

$T_2(x,y) = (2x, y, x+y)$ is $T_2 = \begin{bmatrix} 2 & 0 \\ 0 & 1 \\ 1 & 1 \end{bmatrix}$.

We know that from the above composition formula $T_2 \circ T_1 = T_2 \cdot T_1$

$$= \begin{bmatrix} 2 & 0 \\ 0 & 1 \\ 1 & 1 \end{bmatrix} \begin{bmatrix} 1 & 1 \\ 0 & 1 \end{bmatrix}$$

$$= \begin{bmatrix} 2 & 2 \\ 0 & 1 \\ 1 & 2 \end{bmatrix}$$

Therefore, $T_2 \circ T_1(x,y) = \begin{bmatrix} 2 & 2 \\ 0 & 1 \\ 1 & 2 \end{bmatrix} \begin{bmatrix} x \\ y \end{bmatrix}$

$$T_2 \circ T_1(x,y) = (2x + 2y, \ y, \ x + 2y).$$

Example 1.7

Find the standard matrix for the linear operator $T : R^2 \rightarrow R^2$ that is a dilation of a vector with factor $k = 2$, then the resulting vector is rotated by $\theta = 45°$, and then contraction is applied with factor $k = \sqrt{2}$.

Solution

The linear operator T can be expressed as the composition $T = T_3 \circ T_2 \circ T_1$, where T_1 is the dilation with factor $k = 2$, T_2 is the rotation with $\theta = 45°$, and T_3 is the contraction with factor $k = \sqrt{2}$. If the standard matrices for these linear transformations are

$$T_1 = \begin{bmatrix} k & 0 \\ 0 & k \end{bmatrix} \text{ with } k = 2, \text{ then } T_1 = \begin{bmatrix} 2 & 0 \\ 0 & 2 \end{bmatrix}$$

$$T_2 = \begin{bmatrix} \cos\theta & -\sin\theta \\ \sin\theta & \cos\theta \end{bmatrix} \text{ with } \theta = 45°, \text{ then } T_2 = \begin{bmatrix} \dfrac{1}{\sqrt{2}} & -\dfrac{1}{\sqrt{2}} \\ \dfrac{1}{\sqrt{2}} & \dfrac{1}{\sqrt{2}} \end{bmatrix}$$

$$T_3 = \begin{bmatrix} k & 0 \\ 0 & k \end{bmatrix} \text{ with } k = \sqrt{2}, \text{ then } T_3 = \begin{bmatrix} \sqrt{2} & 0 \\ 0 & \sqrt{2} \end{bmatrix}.$$

Hence, $T = T_3 \circ T_2 \circ T_1 = T_3 \cdot T_2 \cdot T_1 = \begin{bmatrix} 2 & 0 \\ 0 & 2 \end{bmatrix} \cdot \begin{bmatrix} \dfrac{1}{\sqrt{2}} & -\dfrac{1}{\sqrt{2}} \\ \dfrac{1}{\sqrt{2}} & \dfrac{1}{\sqrt{2}} \end{bmatrix} \cdot \begin{bmatrix} \sqrt{2} & 0 \\ 0 & \sqrt{2} \end{bmatrix}.$

$$\therefore T = T_3 \circ T_2 \circ T = \begin{bmatrix} 2 & -2 \\ 2 & 2 \end{bmatrix}.$$

Example 1.8

Find the standard matrix for the linear operator $T : R^3 \rightarrow R^3$ that first reflects the vector about the yz-plane, then the resulting vector rotates counterclockwise about the z-axis through an angle $\theta = 90°$, and then projects that vector orthogonally onto the xy-plane.

Solution

The linear operator T can be expressed as the composition $T = T_3 \circ T_2 \circ T_1$, where T_1 is the reflection about the yz-plane, T_2 is the rotation about z-axis, and T_3 is the orthogonal projection on the xy-plane. The standard matrices for these linear

transformations are $T_1 = \begin{bmatrix} -1 & 0 & 0 \\ 0 & 1 & 0 \\ 0 & 0 & 1 \end{bmatrix}$, $T_2 = \begin{bmatrix} \cos\theta & -\sin\theta & 0 \\ \sin\theta & \cos\theta & 0 \\ 0 & 0 & 1 \end{bmatrix}$ at $\theta = 90°$,

$T_2 = \begin{bmatrix} 0 & -1 & 0 \\ 1 & 0 & 0 \\ 0 & 0 & 1 \end{bmatrix}$, and $T_3 = \begin{bmatrix} 1 & 0 & 0 \\ 0 & 1 & 0 \\ 0 & 0 & 0 \end{bmatrix}.$

Therefore, $T = T_3 \circ T_2 \circ T_1$

$$T = \begin{bmatrix} 1 & 0 & 0 \\ 0 & 1 & 0 \\ 0 & 0 & 0 \end{bmatrix} \cdot \begin{bmatrix} 0 & -1 & 0 \\ 1 & 0 & 0 \\ 0 & 0 & 1 \end{bmatrix} \begin{bmatrix} -1 & 0 & 0 \\ 0 & 1 & 0 \\ 0 & 0 & 1 \end{bmatrix}$$

$$T = \begin{bmatrix} 0 & -1 & 0 \\ -1 & 0 & 0 \\ 0 & 0 & 0 \end{bmatrix}.$$

EXERCISE SET 1

Q.1 Find the standard matrix for the linear transformation T defined by the formula
   a. $T(x_1, x_2) = (x_2, x_1)$
   b. $T(x, y, z, w) = (w, x, z, y, x - z)$
   c. $T(x_1, x_2, x_3) = (0, 0, 0, 0, 0)$

Q.2 Find the standard matrix for the linear operator $T : R^2 \rightarrow R^2$ that is first rotated by $90°$, then the resulting vector is reflected about the line $y = x$.

Q.3 Find the standard matrix for the linear operator $T : R^2 \rightarrow R^2$ that is first an orthogonal projection on the y-axis, then the resulting vector is contracted with factor $k = \dfrac{1}{2}$.

Q.4 Find the standard matrix for the linear operator $T : R^2 \rightarrow R^2$ that is first a reflection about the x-axis, then the resulting vector is dilated with factor $k = 3$.

Q.5 Find the standard matrix for the linear operator $T : R^3 \rightarrow R^3$ that is reflected about the yz-plane followed by an orthogonal projection on the xz-plane.

Q.6 Find the standard matrix for the linear operator $T : R^3 \rightarrow R^3$ that is projected orthogonally on the xy-plane followed by the reflection about the yz-plane.

Q.7 Find the standard matrix for the linear operator $T : R^3 \rightarrow R^3$ that is first rotated by $270°$ about the x-axis, then the resulting vector is a rotation of $90°$ about the y-axis and then a rotation of $180°$ about the z-axis.

Q.8 Determine whether $T_1 \circ T_2 = T_2 \circ T_1$
   a. $T_1 : R^2 \rightarrow R^2$ is the orthogonal projection on the x-axis, and $T_2 : R^2 \rightarrow R^2$ is the orthogonal projection on the y-axis.
   b. $T_1 : R^3 \rightarrow R^3$ is a dilation by a factor k and $T_2 : R^3 \rightarrow R^3$ is the rotation about the z-axis through an angle θ.

ANSWERS TO EXERCISE SET 1

1. a. $T = \begin{bmatrix} 0 & 1 \\ 1 & 0 \end{bmatrix}$ b. $T = \begin{bmatrix} 0 & 0 & 0 & 1 \\ 1 & 0 & 0 & 0 \\ 0 & 0 & 1 & 0 \\ 0 & 1 & 0 & 0 \\ 1 & 0 & -1 & 0 \end{bmatrix}$

c. $T = \begin{bmatrix} 0 & 0 & 0 \\ 0 & 0 & 0 \\ 0 & 0 & 0 \\ 0 & 0 & 0 \\ 0 & 0 & 0 \end{bmatrix}$

2. $T = \begin{bmatrix} 1 & 0 \\ 0 & -1 \end{bmatrix}$

3. $T = \begin{bmatrix} 0 & 0 \\ 0 & \dfrac{1}{2} \end{bmatrix}$

4. $T = \begin{bmatrix} 3 & 0 \\ 0 & -3 \end{bmatrix}$

5. $T = \begin{bmatrix} -1 & 0 & 0 \\ 0 & 0 & 0 \\ 0 & 0 & 1 \end{bmatrix}$

6. $T = \begin{bmatrix} -1 & 0 & 0 \\ 0 & 1 & 0 \\ 0 & 0 & 0 \end{bmatrix}$

7. $T = \begin{bmatrix} 0 & 1 & 0 \\ 0 & 0 & -1 \\ 1 & 0 & 0 \end{bmatrix}$

8. a. yes   b. yes

2 General Linear Transformations

In Chapter 1, we defined linear transformations from R^n to R^m. In this chapter, we shall define and study linear transformations from an arbitrary vector space V to another arbitrary vector space W as are an extension to the linear transformation from R^n to R^m and also, how to formulate the formula of the linear transformation using bases of the vector space. Moreover, kernel space and range of linear transformations will be discussed in this chapter. Bases for the kernel space and range of the linear transformation shall also be derived in this chapter. Then, statement and proof of the dimension theorem for linear transformations will be derived. That is nothing but rank nullity theorem for the vector space. Examples are included in Exercise Set 2.

2.1 INTRODUCTION OF GENERAL LINEAR TRANSFORMATIONS

Definition: If $T : V \to W$ is a function from a vector space V into a vector space W, then T is called a linear transformation from V to W if it satisfies the following properties:

a. Additive property: $T(u + v) = T(u) + T(v), \forall u, v \in V$.
b. Homogeneous property: $T(ku) = kT(u), \forall k \in R, u \in V$.

When we take $V = W$, the linear transformation $T : V \to V$ is called a linear operator on V.

2.1.1 PROPERTIES OF LINEAR TRANSFORMATIONS

If $T : V \to W$ is a linear transformation, then

i. $T(0) = 0$
ii. $T(-v) = -T(v)$ for all v in V
iii. $T(v - w) = T(v) - T(w)$ for all $v, w \in V$
iv. $T(k_1 v_1 + k_2 v_2 + \cdots + k_n v_n) = k_1 T(v_1) + k_2 T(v_2) + \cdots + k_n T(v_n)$,

where $v_1, v_2, \ldots, v_n$ are vectors in V and $k_1, k_2, \ldots, k_n$ are all scalars.

Example 2.1

Determine whether the transformation $T: R^3 \rightarrow R^2$ given by the formula $T(x_1, x_2, x_3) = (2x_1 - x_2 + x_3, x_2 - 4x_3)$ is linear or not.

Solution

Let $u = (x_1, x_2, x_3) \in R^3$, such that $T(u) = (2x_1 - x_2 + x_3, x_2 - 4x_3)$ and $v = \left(x_1', x_2', x_3'\right) \in R^3$, such that $T(v) = \left(2x_1' - x_2' + x_3', x_2' - 4x_3'\right)$.

To prove, $T(u+v) = T(u) + T(v)$

Let $u + v = \left(x_1 + x_1', x_2 + x_2', x_3 + x_3'\right) \in R^3$

$$\text{L.H.S.} = {}'T(u+v) = \left(2\left(x_1 + x_1'\right) - \left(x_2 + x_2'\right) + \left(x_3 + x_3'\right), \ \left(x_2 + x_2'\right) - 4\left(x_3 + x_3'\right)\right)$$

$$= \left(2x_1 - x_2 + x_3 + 2x_1' - x_2' + x_3', \ x_2 - 4x_3 + x_2' - 4x_3'\right)$$

$$= \left(2x_1 - x_2 + x_3, \ x_2 - 4x_3\right) + \left(2x_1' - x_2' + x_3', \ x_2' - 4x_3'\right)$$

$$= T(u) + T(v)$$

$$= \text{R.H.S.}$$

Therefore, property (a) is held.

$$k \in R, \ u = (x_1, x_2, x_3) \in R^3.$$

To prove, $T(ku) = kT(u)$

$$T(ku) = T\left(kx_1, kx_2, kx_3\right) = \left(2kx_1 - kx_2 + kx_3, kx_2 - 4kx_3\right)$$

$$= k\left(2x_1 - kx_2 + kx_3, kx_2 - 4kx_3\right)$$

$$= kT(u)$$

$$= \text{R.H.S.}$$

Therefore, property (b) is also held.

Example 2.2

Determine whether the transformation $T: M_{22} \rightarrow R$, where $T(A) = \det(A)$, is linear or not.

Solution

Let $A = \begin{bmatrix} a & b \\ c & d \end{bmatrix} \in M_{22}$, such that $T(A) = ad - bc$

and $B = \begin{bmatrix} a_1 & b_1 \\ c_1 & d_1 \end{bmatrix} \in M_{22}$, such that $T(B) = a_1 d_1 - b_1 c_1$.

To prove $T(A+B) = T(A) + T(B)$

Let $A + B = \begin{bmatrix} a & b \\ c & d \end{bmatrix} + \begin{bmatrix} a_1 & b_1 \\ c_1 & d_1 \end{bmatrix} = \begin{bmatrix} a+a_1 & b+b_1 \\ c+c_1 & d+d_1 \end{bmatrix}$,

$$\text{L.H.S.} = T(A+B) = (a+a_1)(d+d_1) - (b+b_1)(c+c_1)$$

$$= ad + ad_1 + a_1 d + a_1 d_1 - bc - bc_1 - c_1 b - b_1 c_1$$

$$= ad - bc + a_1 d_1 - b_1 c_1 + ad_1 + a_1 d - bc_1 - b_1 c$$

$$= \det(A) + \det(B) + ad_1 + a_1 d - bc_1 - b_1 c$$

$$\neq T(A) + T(B)$$

$$\neq \text{R.H.S.}$$

Here, property (a) is not held.
Therefore, given transformation is not linear.

Example 2.3

Determine whether the transformation $T : P_2 \to P_2$, where
$T(a_0 + a_1 x + a_2 x^2) = (a_0 + 1) + (a_1 + 1)x + (a_2 + 1)x^2$, is linear or not.

Solution

Let $P(x) = a_0 + a_1 x + a_2 x^2 \in P_2$, such that

$$T(P(x)) = (a_0 + 1) + (a_1 + 1)x + (a_2 + 1)x^2$$

and $Q(x) = b_0 + b_1 x + b_2 x^2 \in P_2$, such that $T(Q(x)) = (b_0 + 1) + (b_1 + 1)x + (b_2 + 1)x^2$.
To prove, $T(P(x) + Q(x)) = T(P(x)) + T(Q(x))$
Let $P(x) + Q(x) = (a_0 + b_0) + (a_1 + b_1)x + (a_2 + b_2)x^2$

$$\text{L.H.S.} = T(P(x) + Q(x)) = (a_0 + b_0 + 1) + (a_1 + b_1 + 1)x + (a_2 + b_2 + 1)x^2$$

$$\text{R.H.S.} = T(P(x)) + T(Q(x))$$

$$= (a_0 + 1) + (a_1 + 1)x + (a_2 + 1)x^2 + (b_0 + 1) + (b_1 + 1)x + (b_2 + 1)x^2$$

$$= (a_0 + b_0 + 2) + (a_1 + b_1 + 2)x + (a_2 + b_2 + 2)x^2$$

From the above calculation, $T(P(x) + Q(x)) \neq T(P(x)) + T(Q(x))$.
Property (a) is not held.
Hence, given transformation is not linear.

Example 2.4

Determine whether the operator $T : F(-\infty, \infty) \to F(-\infty, \infty)$, where $T(f(x)) = f(x+1)$, is linear or not.

Solution

Let $f(x) \in F(-\infty, \infty)$, such that $T(f(x)) = f(x+1)$
and $g(x) \in F(-\infty, \infty)$, such that $T(g(x)) = g(x+1)$.
To prove, $T((f+g)(x)) = T(f(x)) + T(g(x))$.
Let $(f+g)(x) = f(x) + g(x)$

$$\text{L.H.S.} = T((f+g)(x))$$

$$= (f+g)(x+1)$$

$$= (f(x+1) + g(x+1))$$

$$= T(f(x)) + T(g(x))$$

$$= \text{R.H.S.}$$

Property (a) is held.

Theorem 2.1

If $T : R^n \to R^m$ is a linear transformation and $e_1, e_2, \ldots, e_n$ are the standard basis vectors for R^n, then the standard matrix for T is $[T] = [T(e_1) \mid T(e_2) \mid \ldots \mid T(e_n)]$.

The above formula is very powerful for obtaining standard matrices and evaluating the effect of the linear transformation.

2.2 FORMING A LINEAR TRANSFORMATION

2.2.1 OBTAINING LINEAR TRANSFORMATIONS FROM BASIS VECTORS AND THEIR IMAGE VECTORS

A more general result of theorem 2.1 is, $T : V \to W$ is a linear transformation, and if $\{v_1, v_2, v_3, \ldots, v_n\}$ is any basis for V, then the image $T(v)$ of any vector v in V can be calculated from the images $T(v_1), T(v_2), \ldots, T(v_n)$ of the basis vectors. This can be obtained by first expressing v as a linear combination of the basis vectors, i.e., $v = k_1 v_1 + k_2 v_2 + \cdots + k_n v_n$, and then applying transformation on both sides, we get

$$T(v) = k_1 T(v_1) + k_2 T(v_2) + \cdots + k_n T(v_n).$$

Example 2.5

Consider the basis $S = \{v_1, v_2, v_3\}$ for R^3, where $v_1 = (1, 1, 1)$, $v_2 = (1, 1, 0)$, and $v_3 = (1, 0, 0)$. Let $T : R^3 \to R^2$ be the linear transformation such that $T(v_1) = (1, 0)$,

$T(v_2) = (2, -1)$, and $T(v_3) = (4, 3)$. Find a formula for $T(x_1, x_2, x_3)$; then, use this formula to compute $T(2, -3, 5)$.

Solution

Suppose $(x_1, x_2, x_3) \in R^3$ as a linear combination of $v_1 = (1, 1, 1)$, $v_2 = (1, 1, 0)$, and $v_3 = (1, 0, 0)$.

Therefore, by definition of linear combination,

$$\therefore (x_1, x_2, x_3) = k_1 v_1 + k_2 v_2 + k_3 v_3$$

$$\therefore (x_1, x_2, x_3) = k_1 (1, 1, 1) + k_2 (1, 1, 0) + k_3 (1, 0, 0).$$

$$\therefore (x_1, x_2, x_3) = (k_1 + k_2 + k_3, k_1 + k_2, k_1)$$

Comparing both sides, we get

$$k_1 = x_3, \quad k_1 + k_2 = x_2, \quad k_1 + k_2 + k_3 = x_1$$

$$\therefore k_1 = x_3, \quad k_2 = x_2 - x_3, \quad k_3 = x_1 - x_2$$

$$T(x_1, x_2, x_3) = x_3 (1, 0) + (x_2 - x_3)(2, -1) + (x_1 - x_2)(4, 3)$$

$$= (x_3 + 2(x_2 - x_3) + 4(x_1 - x_2), 0 - 1(x_2 - x_3) + 3(x_1 - x_2))$$

$$= (x_3 + 2x_2 - 2x_3 + 4x_1 - 4x_2, -x_2 + x_3 + 3x_1 - 3x_2)$$

$$= (4x_1 - 2x_2 - x_3, 3x_1 - 4x_2 + x_3)$$

From this formula, we obtain

$$T(2, -3, 5) = (9, 23).$$

Example 2.6

Consider the basis $S = \{v_1, v_2\}$ for R^2, where $v_1 = (-2, 1)$ and $v_2 = (1, 3)$. Let $T : R^2 \rightarrow R^3$ be the linear transformation such that $T(v_1) = (-1, 2, 0)$ and $T(v_2) = (0, -3, 5)$. Find a formula for $T(x_1, x_2)$; then, use this formula to compute $T(2, -3)$.

Solution

Suppose $(x_1, x_2) \in R^2$ as a linear combination of $v_1 = (-2, 1)$ and $v_2 = (1, 3)$. Therefore, by definition of linear combination,

$$\therefore (x_1, x_2) = k_1 v_1 + k_2 v_2$$

$$\therefore (x_1, x_2) = k_1 (-2, 1) + k_2 (1, 3)$$

$$\therefore (x_1, x_2) = (-2k_1 + k_2, k_1 + 3k_2).$$

Comparing both sides, we get

$$-2k_1 + k_2 = x_1, \quad k_1 + 3k_2 = x_2$$

$$\therefore k_1 = \frac{x_2 - 3x_1}{7}, \quad k_2 = \frac{x_1 + 2x_2}{7}$$

$$T(x_1, x_2) = \left(\frac{x_2 - 3x_1}{7}\right) T(-2, 1) + T(1, 3)\left(\frac{x_1 + 2x_2}{7}\right)$$

$$= \left(\frac{x_2 - 3x_1}{7}\right)(-1, 2, 0) + \left(\frac{x_1 + 2x_2}{7}\right)(0, -3, 5)$$

$$= \left(\frac{-x_2 + 3x_1}{7}, \frac{-9x_1 - 4x_2}{7}, \frac{5x_1 + 10x_2}{7}\right).$$

From this formula, we obtain

$$T(2,-3) = \left(\frac{9}{7}, \frac{-6}{7}, \frac{-20}{7}\right).$$

Example 2.7

Consider the basis $S = \{v_1, v_2\}$ for R^2, where $v_1 = \begin{bmatrix} 1 \\ 1 \end{bmatrix}$ and $v_2 = \begin{bmatrix} 2 \\ 3 \end{bmatrix}$ and let $T: R^2 \to P_2$ be the linear transformation such that $T(v_1) = 2 - 3x + x^2$ and $T(v_2) = 1 - x^2$. Find $T\begin{bmatrix} a \\ b \end{bmatrix}$ and then find $T\begin{bmatrix} -1 \\ 2 \end{bmatrix}$.

Solution

Suppose $(a,b) \in R^2$ as a linear combination of $v_1 = \begin{bmatrix} 1 \\ 1 \end{bmatrix}$ and $v_2 = \begin{bmatrix} 2 \\ 3 \end{bmatrix}$.

Therefore, by definition of linear combination,

$$\therefore \begin{bmatrix} a \\ b \end{bmatrix} = k_1 \begin{bmatrix} 1 \\ 1 \end{bmatrix} + k_2 \begin{bmatrix} 2 \\ 3 \end{bmatrix}$$

$$\therefore \begin{bmatrix} a \\ b \end{bmatrix} = \begin{bmatrix} k_1 + 2k_2 \\ k_1 + 3k_2 \end{bmatrix}.$$

Comparing both sides, we get

$$k_1 + 2k_2 = a, \quad k_1 + 3k_2 = b$$

$$\therefore k_1 = 3a - 2b, \quad k_2 = b - a$$

$$T\begin{bmatrix} a \\ b \end{bmatrix} = (3a - 2b)T\begin{bmatrix} 1 \\ 1 \end{bmatrix} + (b - a)T\begin{bmatrix} 2 \\ 3 \end{bmatrix}$$

$$= (3a - 2b)(2 - 3x + x^2) + (b - a)(1 - x^2)$$

$$= (4a - 3b)x^2 + (-9a + 6b)x + (5a - 3b).$$

From this formula, we obtain

$$T\begin{bmatrix} a \\ b \end{bmatrix} = -11 + 21x - 10x^2.$$

Example 2.8

Let $T: M_{22} \to R$ be a linear transformation where $T(v_1) = 1$, $T(v_2) = 2$, $T(v_3) = 3$, $T(v_4) = 4$, where $v_1 = \begin{bmatrix} 1 & 0 \\ 0 & 0 \end{bmatrix}$, $v_2 = \begin{bmatrix} 1 & 1 \\ 0 & 0 \end{bmatrix}$, $v_3 = \begin{bmatrix} 1 & 1 \\ 1 & 0 \end{bmatrix}$, and $v_4 = \begin{bmatrix} 1 & 1 \\ 1 & 1 \end{bmatrix}$. Find $T\begin{bmatrix} a & b \\ c & d \end{bmatrix}$ and $T\begin{bmatrix} 1 & 3 \\ 4 & 2 \end{bmatrix}$.

Solution

Let $v = \begin{bmatrix} a & b \\ c & d \end{bmatrix}$ be an arbitrary vector in M_{22} and can be expressed as a linear combination of v_1, v_2, v_3 and v_4.

$$\therefore \begin{bmatrix} a & b \\ c & d \end{bmatrix} = k_1 v_1 + k_2 v_2 + k_3 v_3 + k_4 v_4$$

$$\therefore \begin{bmatrix} a & b \\ c & d \end{bmatrix} = k_1 \begin{bmatrix} 1 & 0 \\ 0 & 0 \end{bmatrix} + k_2 \begin{bmatrix} 1 & 1 \\ 0 & 0 \end{bmatrix} + k_3 \begin{bmatrix} 1 & 1 \\ 1 & 0 \end{bmatrix} + k_4 \begin{bmatrix} 1 & 1 \\ 1 & 1 \end{bmatrix}$$

$$\therefore \begin{bmatrix} a & b \\ c & d \end{bmatrix} = \begin{bmatrix} k_1 + k_2 + k_3 + k_4 & k_2 + k_3 + k_4 \\ k_3 + k_4 & k_4 \end{bmatrix}.$$

Comparing both sides, we get

$$k_1 + k_2 + k_3 + k_4 = a, k_2 + k_3 + k_4 = b, k_3 + k_4 = c \text{ and } k_4 = d.$$

Solving these equations

$$\therefore k_1 = a - b, \quad k_2 = b - c, \quad k_3 = c - d, \quad k_4 = d$$

$$\therefore T \begin{bmatrix} a & b \\ c & d \end{bmatrix} = k_1 T(v_1) + k_2 T(v_2) + k_3 T(v_3) + k_4 T(v_4)$$

$$= (a-b)(1) + (b-c)(2) + (c-d)(3) + d(4)$$

$$\therefore T \begin{bmatrix} a & b \\ c & d \end{bmatrix} = a - b + 2b - 2c + 3c - 3d + 4d$$

$$= a + b + c + d$$

$$\therefore T \begin{bmatrix} 1 & 3 \\ 4 & 2 \end{bmatrix} = 1 + 3 + 4 + 2 = 10.$$

2.3   KERNEL AND RANGE OF LINEAR TRANSFORMATION T

In this section, we will define kernel space and range of linear transformation T.

Definition: Kernel of Linear Transformation T

If $T : V \rightarrow W$ is a linear transformation, then the set of vectors in V that T maps into 0 is called the kernel of T. It is denoted by $\ker(T)$.

Definition: Range of Linear Transformation T

The set of all vectors in W that are images under T of at least one vector in V is called the range of T. It is denoted by $R(T)$.

Theorem 2.2

If $T : V \rightarrow W$ is a linear transformation, then (i) the kernel of T is a subspace of V and (ii) the range of T is a subspace of W.

Remark: The kernel of T_A is the null space of A, and the range of T_A is the column space of A.

Definition: Rank of T

If $T : V \rightarrow W$ is a linear transformation, then the dimension of the range of T is called the rank of T and is denoted by $\text{rank}(T)$.

Definition: Nullity of T

The dimension of the kernel is called the nullity of T and is denoted by $\text{nullity}(T)$.

Theorem 2.3: Dimension Theorem for Transformation[4]

If $T : V \to W$ is a linear transformation, then $R(T) + \ker(T) = \dim(V)$, where n is the number of column of T_A.

Proof[4]: Let $T : V \to W$ is a linear transformation

Here, there are three cases that are taken into consideration.

Case 1: If $\ker(T) = 0$, then nullity $(T) = 0$

Firstly, we have to show that the vectors $T(v_1), T(v_2), \ldots, T(v_n)$, where $v_1, v_2, \ldots, v_n$ are the basis for V and therefore, $v_1, v_2, \ldots, v_n$ are linearly independent.

Suppose $k_1 T(v_1) + k_2 T(v_2) + \cdots + k_n T(v_n) = 0$, where $k_1, k_2, \ldots, k_n \in R$.

Then, $T\big(k_1(v_1) + k_2(v_2) + \cdots + k_n(v_n)\big) = 0$ (by linearity property)

$\big(k_1(v_1) + k_2(v_2) + \cdots + k_n(v_n)\big) = 0 \, \big(\because \ker(T) = 0\big)$

$v_1 = 0, v_2 = 0, \ldots, v_n = 0 \, \big(\because v_i \text{ are LI}\big)$

Hence, $\mathrm{Im}(T) = \langle w_1, w_2, \ldots, w_n \rangle$.

So, $\mathrm{rank}(T) + \mathrm{nullity}(T) = \dim(\mathrm{Im}\,T) + 0 = n = \dim(V)$.

Case 2: If $\ker(T) = V$, then nullity $(T) = \dim(V)$

Hence, $\mathrm{Im}(T) = \{0\} \Rightarrow \mathrm{rank}(T) = 0$.

Therefore, $\mathrm{rank}(T) + \mathrm{nullity}(T) = 0 + \dim(V) = \dim(V)$.

Case 3: $0 < \mathrm{nullity}(T) < \dim(V)$

Let $v_1, v_2, \ldots, v_r$ be a basis for $\ker(T)$ and $n = \dim(V)$, so $r = \mathrm{nullity}\,T$ and $r < n$.

Extend the basis $v_1, v_2, \ldots, v_r$ to form a basis $v_1, v_2, \ldots, v_r, v_{r+1}, \ldots, v_n$ of V.

Then, $T(v_{r+1}), \ldots, T(v_n)$ spam $\mathrm{Im}(W)$. For

$\mathrm{Im}(T) = \langle T(v_1), T(v_2), \ldots, T(v_r), T(v_{r+1}), \ldots, T(v_n) \rangle$

$$= \langle 0, 0, \ldots, 0, T(v_{r+1}), \ldots, T(v_n) \rangle$$

$$= \langle T(v_{r+1}), \ldots, T(v_n) \rangle$$

So assume

$$x_1 T(v_{r+1}) + \cdots + x_{n-r} T(v_n) = 0$$

$\Rightarrow \qquad T(x_1 v_{r+1} + \cdots + x_{n-r} v_n) = 0$

$\Rightarrow \qquad x_1 v_{r+1} + \cdots + x_{n-r} v_n \in \ker(T)$

$\Rightarrow \qquad x_1 v_{r+1} + \cdots + x_{n-r} v_n = y_1 v_1 + \cdots + y_r v_r$ for some $y_1, y_2, \ldots, y_r$

$\Rightarrow \qquad (-y_1) v_1 + \cdots + (-y_r) v_r + x_1 v_{r+1} + \cdots + x_{n-r} v_n = 0$

And since $v_1, v_2, \ldots, v_r$ is the basis for V and all coefficients are zero.

Therefore, $\mathrm{rank}(T) + \mathrm{nullity}(T) = (n - r) + r = \dim(V)$.

Hence, the proof.

Example 2.9

Verify the dimension theorem for the linear operator $T: R^2 \to R^2$ which is given by the formula $T(x,y)=(2x-y,-8x+4y)$.

Solution

Let the standard matrix for the linear operator $T(x,y)=(2x-y,-8x+4y)$ is

$$T = \begin{bmatrix} 2 & -1 \\ -8 & 4 \end{bmatrix}.$$

The solution of $T(X)=0$ by Gauss elimination is

$$\text{Augmented matrix} = \begin{bmatrix} 2 & -1 & 0 \\ -8 & 4 & 0 \end{bmatrix}.$$

By taking $\frac{1}{2}R_1$,

$$\approx \begin{bmatrix} 1 & -\frac{1}{2} & 0 \\ -8 & 4 & 0 \end{bmatrix}$$

$$R_2 \to R_2 + 8R_1$$

$$\approx \begin{bmatrix} 1 & -\frac{1}{2} & 0 \\ 0 & 0 & 0 \end{bmatrix}.$$

By taking back substitution,

$$x - \frac{1}{2}y = 0.$$

Here, there are two unknowns and one equation; hence, we have to choose one variable as a parameter.

Take $y \in R$

$$\therefore \text{solution set} = \left\{ (x,y) = t\left(\frac{1}{2},1\right) / t \in R \right\}.$$

We know that the basis for the column space is equal to basic for the $R(T)$.

Therefore, basis for the $R(T) = \left\{ \begin{pmatrix} 2 \\ -8 \end{pmatrix} \right\}.$

Hence, $\dim(R(T))=1$.

The basis for the null space is $\left(\dfrac{1}{2}, 1\right)$; hence, the basis for $\ker(T) = \left\{\begin{pmatrix} \dfrac{1}{2} \\ 1 \end{pmatrix}\right\}$.

Therefore, $\dim(\ker(T)) = 1$.

As a result, $\dim(R(T)) + \dim(\ker(T)) = 1 + 1 = 2 = \dim(R^2)$.

Hence, it is verified.

Example 2.10

Describe the kernel and range of

    a. The orthogonal projection on the xz-plane.
    b. The orthogonal projection on the yz-plane.
    c. The orthogonal projection on the plane defined by the equation $y = x$.

Solution

    a. For the orthogonal projection on the xz-plane, the formula for transformation is $T(x, y, z) = (x, 0, z)$.

        Now, the standard matrix for the above transformation is

$$T = \begin{bmatrix} 1 & 0 & 0 \\ 0 & 0 & 0 \\ 0 & 0 & 1 \end{bmatrix}.$$

The solution of the $T(X) = 0$ by Gauss elimination is

$$\text{Augmented matrix} = \left[\begin{array}{ccc|c} 1 & 0 & 0 & 0 \\ 0 & 0 & 0 & 0 \\ 0 & 0 & 1 & 0 \end{array}\right].$$

Take $R_2 \leftrightarrow R_3$

$$\approx \left[\begin{array}{ccc|c} 1 & 0 & 0 & 0 \\ 0 & 0 & 1 & 0 \\ 0 & 0 & 0 & 0 \end{array}\right].$$

By taking back substitution,
    $x = 0$, $z = 0$, and y is an arbitrary variable.
    Therefore, $y = t$, $t \in R$

$$\text{Solution set} = \left\{(x, y, z) = t(0, 1, 0) / t \in R\right\}.$$

Therefore, $\text{Ker}(T) = \{(x, y, z) = t(0, 1, 0) / t \in R\}$.

And we know that the column space of the matrix T is equal to $R(T)$.

Therefore, $R(T) = \left\{ t_1 \begin{pmatrix} 1 \\ 0 \\ 0 \end{pmatrix} + t_2 \begin{pmatrix} 0 \\ 0 \\ 1 \end{pmatrix} \right\}$.

b. For the orthogonal projection on the yz-plane, the formula for transformation is $T(x, y, z) = (0, y, z)$.

Now, the standard matrix for the above transformation is

$$T = \begin{bmatrix} 0 & 0 & 0 \\ 0 & 1 & 0 \\ 0 & 0 & 1 \end{bmatrix}.$$

The solution of the $T(X) = 0$ by Gauss elimination is

$$\text{Augmented matrix} = \begin{bmatrix} 0 & 0 & 0 & | & 0 \\ 0 & 1 & 0 & | & 0 \\ 0 & 0 & 1 & | & 0 \end{bmatrix}.$$

Take $R_1 \leftrightarrow R_2$

$$\approx \begin{bmatrix} 0 & 1 & 0 & | & 0 \\ 0 & 0 & 0 & | & 0 \\ 0 & 0 & 1 & | & 0 \end{bmatrix}$$

$$R_2 \leftrightarrow R_3$$

$$\approx \begin{bmatrix} 0 & 1 & 0 & | & 0 \\ 0 & 0 & 1 & | & 0 \\ 0 & 0 & 0 & | & 0 \end{bmatrix}.$$

By taking back substitution,

$y = 0$, $z = 0$, and x is an arbitrary variable.

Therefore, $x = t$, $t \in R$

$$\text{Solution set} = \{(x, y, z) = t(1, 0, 0) / t \in R\}.$$

Therefore, $\text{Ker}(T) = \{(x, y, z) = t(1, 0, 0) / t \in R\}$.

And we know that the column space of the matrix T is equal to $R(T)$.

Therefore, $R(T) = \left\{ t_1 \begin{pmatrix} 0 \\ 1 \\ 0 \end{pmatrix} + t_2 \begin{pmatrix} 0 \\ 0 \\ 1 \end{pmatrix} \right\}$.

c. For the orthogonal projection on the plane defined by the equation $y = x$, the formula for transformation is $T(x,y,z) = (x,x,z)$.

Now, the standard matrix for the above transformation is

$$T = \begin{bmatrix} 1 & 0 & 0 \\ 1 & 0 & 0 \\ 0 & 0 & 1 \end{bmatrix}.$$

The solution of the $T(X) = 0$ by Gauss elimination is

$$\text{Augmented matrix} = \begin{bmatrix} 1 & 0 & 0 & | & 0 \\ 1 & 0 & 0 & | & 0 \\ 0 & 0 & 1 & | & 0 \end{bmatrix}.$$

Take $R_1 \rightarrow R_2 - R_1$

$$\approx \begin{bmatrix} 1 & 0 & 0 & | & 0 \\ 0 & 0 & 0 & | & 0 \\ 0 & 0 & 1 & | & 0 \end{bmatrix}$$

$$R_2 \leftrightarrow R_3$$

$$\approx \begin{bmatrix} 1 & 0 & 0 & | & 0 \\ 0 & 0 & 1 & | & 0 \\ 0 & 0 & 0 & | & 0 \end{bmatrix}.$$

By taking back substitution,
$x = 0, z = 0$, and y is an arbitrary variable.
Therefore, $y = t, \quad t \in R$.

$$\text{Solution set} = \left\{ (x,\, y,\, z) = t(0, 1, 0) / t \in R \right\}.$$

Therefore, $\text{Ker}(T) = \left\{ (x,\, y,\, z) = t(0, 1, 0) / t \in R \right\}$.
And we know that the column space of the matrix T is equal to $R(T)$.

$$\text{Therefore, } R(T) = \left\{ t_1 \begin{pmatrix} 1 \\ 1 \\ 0 \end{pmatrix} + t_2 \begin{pmatrix} 0 \\ 0 \\ 1 \end{pmatrix} \right\}.$$

Example 2.11

Verify the dimension theorem for the linear transformation $T : R^4 \rightarrow R^3$ which is given by the formula $T(x_1, x_2, x_3, x_4) = (4x_1 + x_2 - 2x_3 - 3x_4, 2x_1 + x_2 + x_3 - 4x_4, 6x_1 - 9x_3 + 9x_4)$.

Solution

Let the standard matrix for the linear transformations
$$T(x_1,x_2,x_3,x_4)=(4x_1+x_2-2x_3-3x_4,2x_1+x_2+x_3-4x_4,6x_1-9x_3+9x_4)\text{ is}$$

$$T = \begin{bmatrix} 4 & 1 & -2 & -3 \\ 2 & 1 & 1 & -4 \\ 6 & 0 & -9 & 9 \end{bmatrix}.$$

The solution of $T(X)=0$ by Gauss elimination is

$$\text{Augmented matrix} = \left[\begin{array}{cccc|c} 4 & 1 & -2 & -3 & 0 \\ 2 & 1 & 1 & -4 & 0 \\ 6 & 0 & -9 & 9 & 0 \end{array}\right].$$

Take $R_2 - \dfrac{1}{2}R_1,\ \ R_3 - \dfrac{3}{2}R_1,$

$$\approx \left[\begin{array}{cccc|c} 4 & 1 & -2 & -3 & 0 \\ 0 & \frac{1}{2} & 2 & -\frac{5}{2} & 0 \\ 0 & -\frac{3}{2} & -6 & \frac{27}{2} & 0 \end{array}\right]$$

$$\approx \left[\begin{array}{cccc|c} 4 & 1 & -2 & -3 & 0 \\ 0 & 1 & 4 & -5 & 0 \\ 0 & -3 & -12 & 27 & 0 \end{array}\right]$$

$$R_3 + 3R_2$$

$$\approx \left[\begin{array}{cccc|c} 4 & 1 & -2 & -3 & 0 \\ 0 & 1 & 4 & -5 & 0 \\ 0 & 0 & 0 & 12 & 0 \end{array}\right]$$

$$\frac{1}{12}R_3$$

$$\approx \left[\begin{array}{cccc|c} 4 & 1 & -2 & -3 & 0 \\ 0 & 1 & 4 & -5 & 0 \\ 0 & 0 & 0 & 1 & 0 \end{array}\right].$$

By taking back substitution,

$$w = 0,\ y+4z-5w = 0,\ 4x+y-2z-3w = 0$$

$$x = \frac{3}{2}t, \, y = -4t, \, z = t, \, w = 0, \, t \in R$$

$$\text{Solution set} = \left\{ (x,y,z,w) = t\left(\frac{3}{2},-4,1,0\right) \middle/ t \in R \right\}.$$

We know that the basis for the column space is equal to basic for $R(T)$.

$$\text{Therefore, basis for } R(T) = \left\{ \begin{pmatrix} 4 \\ 2 \\ 6 \end{pmatrix}, \begin{pmatrix} 1 \\ 1 \\ 0 \end{pmatrix}, \begin{pmatrix} -3 \\ -4 \\ 9 \end{pmatrix} \right\}.$$

Hence, $\dim(R(T)) = 3$.

The basis for the null space is $\left(\frac{3}{2},-4,1,0\right)$; hence, the basis for $\ker(T) = \left\{ \begin{pmatrix} \frac{3}{2} \\ -4 \\ 1 \\ 0 \end{pmatrix} \right\}.$

Therefore, $\dim(\ker(T)) = 1$.
As a result, $\dim(R(T)) + \dim(\ker(T)) = 3 + 1 = 4 = \dim(R^4)$.
Hence, it is verified.

Example 2.12

Let $T : P_3 \rightarrow P_2$ be the linear transformation defined by $T(ax^3 + bx^2 + cx + d) = 3ax^2 + 2bx + c$. Find $R(T)$ and $\ker(T)$, and verify the dimension theorem.

Solution

We have $T(ax^3 + bx^2 + cx + d) = 3ax^2 + 2bx + c$.

Let $(Ax^2 + Bx + c) \in P_2$ be some element.

$$\therefore Ax^2 + Bx + c = 3ax^2 + 2bx + c$$

$$\Rightarrow A = 3a, \, B = 2b, \, C = c$$

$$\Rightarrow a = \frac{A}{3}, \, b = \frac{B}{2}, \, C = c.$$

Therefore, A, B, and C possess any real value, for which $Ax^2 + Bx + c$ is the image of some vector in P_3.

$$\therefore \text{range of } T = P_2 = 2 + 1 = 3.$$

Now, for the kernel of T,

$$T(ax^3 + bx^2 + cx + d) = \bar{0}, \quad \bar{0} \in P_2.$$

$$\therefore 3ax^2 + 2bx + c = 0x^2 + 0x + 0.$$

$\therefore a = 0, b = 0, c = 0$, and remains arbitrary.

$$\ker(T) = \left\{ 0x^3 + 0x^2 + 0x + d \mid d \in R \right\} = \text{set of all constant polynomial in } P_3.$$

$$\therefore \dim\big(\ker(T)\big) = 1.$$

Dimension theorem: $\dim\big(R(T)\big) + \dim\big(\ker(T)\big) = 3 + 1 = 4 = \dim(P_3)$.
Hence, the proof.

2.4 COMPOSITION OF THE LINEAR TRANSFORMATION

In Chapter 1, we discussed the composition of two or more linear transformations from $R^n \to R^m$. Here, we discuss composition from an arbitrary vector space V to another arbitrary vector space W as an extension to the composition of linear transformation from R^n to R^m. Let $T_1 : V \to U$ and $T_2 : U \to W$ be linear transformations. The application of T_1 followed by-products of a transformation from V to W. This transformation is called the composition of T_2 with T_1 and is denoted by $T_2 \circ T_1$

$$(T_2 \circ T_1)(v) = T_2\big(T_1(v)\big),$$

where v is a vector in V.

Theorem 2.4

If $T : V \to U$ and $S : U \to W$ are linear transformations, then $S \circ T : V \to W$ is also a linear transformation.

Proof: If v_1 and v_2 are the vectors in V and $k \in R$, then by definition of the linear transformation

$$
\begin{aligned}
(S \circ T)(v_1 + v_2) &= S\big(T(v_1 + v_2)\big) \\
&= S\big(T(v_1) + T(v_2)\big) \quad (\because \text{by linearity property}) \\
&= S\big(T(v_1)\big) + S\big(T(v_2)\big) \\
&= S \circ T(v_1) + S \circ T(v_2).
\end{aligned}
$$

First property is held

$$
\begin{aligned}
(S \circ T)(kv_1) &= S\big(T(kv_1)\big) \\
&= S\big(kT(v_1)\big) \\
&= kS\big(T(v_1)\big) \\
&= k\big(S \circ T(v_1)\big).
\end{aligned}
$$

Second property is held.

Thus, $S \circ T : V \to W$ is also a linear transformation.

Example 2.13

Find the domain and co-domain of $T_3 \circ T_2 \circ T_1$ and $(T_3 \circ T_2 \circ T_1)(x, y)$, where $T_1(x, y) = (x + y, y, -x)$, $T_2(x, y, z) = (0, x + y + z, 3y)$, $T_3(x, y, z) = (3x + 2y, 4z - x - 3y)$.

Solution

$T_1(x, y) = (x + y, y, -x)$.

$\quad T_1 : R^2 \to R^3$ is a linear transformation from R^2 to R^3 and also the standard matrix

of $T_1 = \begin{bmatrix} 1 & 1 \\ 0 & 1 \\ -1 & 0 \end{bmatrix}$

$$T_2(x, y, z) = (0, x + y + z, 3y).$$

$T_2 : R^3 \to R^3$ is a linear transformation from R^3 to R^3 and also the standard matrix

of $T_2 = \begin{bmatrix} 0 & 0 & 0 \\ 1 & 1 & 1 \\ 0 & 3 & 0 \end{bmatrix}$

$$T_3(x, y, z) = (3x + 2y, 4z - x - 3y).$$

$T_3 : R^3 \to R^2$ is a linear transformation from R^3 to R^2 and also the standard matrix

of $T_3 = \begin{bmatrix} 3 & 2 & 0 \\ -1 & -3 & 4 \end{bmatrix}$.

Hence, $T_3 \circ T_2 \circ T_1$ is a linear transformation from R^2 to R^2.

Domain of $T_3 \circ T_2 \circ T_1 = R^2$.

Co-domain of $T_3 \circ T_2 \circ T_1 = R^2$.

$$[T_3 \circ T_2 \circ T_1] = \begin{bmatrix} 3 & 2 & 0 \\ -1 & -3 & 4 \end{bmatrix} \begin{bmatrix} 0 & 0 & 0 \\ 1 & 1 & 1 \\ 0 & 3 & 0 \end{bmatrix} \begin{bmatrix} 1 & 1 \\ 0 & 1 \\ -1 & 0 \end{bmatrix}$$

$$[T_3 \circ T_2 \circ T_1] = \begin{bmatrix} 0 & 4 \\ 0 & 6 \end{bmatrix}$$

$$(T_3 \circ T_2 \circ T_1)(x, y) = (4y, 6y).$$

Example 2.14

Let $T_1 : M_{22} \to R$ and $T_2 : M_{22} \to M_{22}$ be the linear transformations given by $T_1(A) = \text{tr}(A)$ and $T_2(A) = A^T$. Find $(T_1 \circ T_2)(A)$, where $T = \begin{bmatrix} a & b \\ c & d \end{bmatrix}$.

Solution

$$T_1(A) = \text{tr}(A) = \text{tr}\begin{bmatrix} a & b \\ c & d \end{bmatrix} = a + d.$$

$$T_2(A) = A^T = \begin{bmatrix} a & b \\ c & d \end{bmatrix}^T = \begin{bmatrix} a & c \\ b & d \end{bmatrix}.$$

$$(T_1 \circ T_2)(A) = T_1(T_2(A)) = T_1(A^T) = tr(A^T) = a + d.$$

Example 2.15

Let $T_1 : P_2 \to P_2$ and $T_2 : P_2 \to P_2$ be the linear transformations given by $T_1(p(x)) = p(x+1)$ and $T_2(p(x)) = xp(x)$. Find $(T_2 \circ T_1)(a_0 + a_1 x + a_2 x^2)$.

Solution

$T_1(p(x)) = p(x+1)$, $T_2(p(x)) = xp(x)$.

$$(T_2 \circ T_1)(a_0 + a_1 x + a_2 x^2) = T_2(T_1(a_0 + a_1 x + a_2 x^2))$$

$$= T_2(a_0 + a_1(x+1) + a_2(x+1)^2)$$

$$= x(a_0 + a_1(x+1) + a_2(x^2 + 2x + 1))$$

$$= (a_0 + a_1 + a_2)x + (a_1 + 2a_2)x^2 + a_2 x^3$$

$$\therefore (T_2 \circ T_1)(a_0 + a_1 x + a_2 x^2) = (a_0 + a_1 + a_2)x + (a_1 + 2a_2)x^2 + a_2 x^3.$$

EXERCISE SET 2

Q.1 **Fill in the blanks**

a. If $T_A : R^n \to R^m$ is multiplied by A, then the null space of A corresponds to the _____ of T_A and the column space of A corresponds to the _____.

b. If V is a finite-dimensional vector space and $T : V \to W$ is a linear transformation, then the dimension of the range of T plus the dimension of the kernel of T is _____.

c. If $T_A : R^5 \to R^3$ is multiplied by A, and if $rank(T_A) = 2$, then the general solution $AX = 0$ has _____ (how many?) parameters.

Q.2   Which of the following are linear transformations? Justify.

   i. $T: R^2 \rightarrow R^2$, where $T(x,y) = (xy,x)$

   ii. $T: R^2 \rightarrow R^3$, where $T(x,y) = (x+1, 2y, x+y)$

   iii. $T: R^3 \rightarrow R^2$, where $T(x,y,z) = (|x|, 0)$

   iv. $T: M_{22} \rightarrow R$, where $T\left(\begin{bmatrix} a & b \\ c & d \end{bmatrix}\right) = 3a - 4b + c - d$

   v. $T: M_{nn} \rightarrow R$, where $T(A) = tr(A)$

Q.3   Consider the basis $S = \{v_1, v_2, v_3\}$ for R^3, where $v_1 = (1,1,1)$, $v_2 = (1,1,0)$, $v_3 = (1,0,0)$ and let $T: R^3 \rightarrow R^3$ be the linear operator such that $T(v_1) = (2,-1,4)$, $T(v_2) = (3,0,1)$, $T(v_3) = (-1,5,1)$. Find a formula for $T(x_1, x_2, x_3)$, and use that formula to find $T(2,4,-1)$.

Q.4   Let $T: M_{22} \rightarrow R$ be a linear transformation for which $T(v_1) = 1$, $T(v_2) = 2$, $T(v_3) = 3$, and $T(v_4) = 4$, where $v_1 = \begin{bmatrix} 1 & 0 \\ 0 & 0 \end{bmatrix}$, $v_2 = \begin{bmatrix} 1 & 1 \\ 0 & 0 \end{bmatrix}$, $v_3 = \begin{bmatrix} 1 & 1 \\ 1 & 0 \end{bmatrix}$, and $v_4 = \begin{bmatrix} 1 & 1 \\ 1 & 1 \end{bmatrix}$. Find a formula for $T\begin{bmatrix} a & b \\ c & d \end{bmatrix}$, and use that formula to find $T\begin{bmatrix} 1 & 0 \\ -3 & 4 \end{bmatrix}$

Q.5   Let $T_1: P_1 \rightarrow P_2$ and $T_2: P_2 \rightarrow P_2$ be the linear transformations given by $T_1(p(x)) = xp(x)$ and $T_2(p(x)) = p(2x+4)$. Find $(T_2 \circ T_1)(a_0 + a_1 x)$.

Q.6   Let $T_1: R^2 \rightarrow M_{22}$ and $T_2: R^2 \rightarrow R^2$ be the linear transformations given by $T_1\begin{bmatrix} a \\ b \end{bmatrix} = \begin{bmatrix} a+b & b \\ 0 & a-b \end{bmatrix}$ and $T_2\begin{bmatrix} c \\ d \end{bmatrix} = \begin{bmatrix} 2c+d \\ -d \end{bmatrix}$. Find $(T_1 \circ T_2)\begin{bmatrix} 2 \\ 1 \end{bmatrix}$ and $(T_1 \circ T_2)\begin{bmatrix} x \\ y \end{bmatrix}$.

Q.7   Let V be any vector space, and let $T: V \rightarrow V$ be defined by $T(\bar{v}) = 3\bar{v}$, then (a) what is the kernel of T? (b) What is the range of T?

Q.8   Let $T: R^2 \rightarrow R^2$ be the linear transformation defined by $T(x,y) = (x,0)$.
   i.   Which of the following vectors are in $\ker(T)$?
       (a) $(0,2)$ (b) $(2,2)$
   ii. Which of the following vectors are in $R(T)$?
       (a) $(3,0)$ (b) $(3,2)$

Q.9   Let $T: P_2 \rightarrow P_3$ be the linear transformation defined by $T(p(x)) = xp(x)$
   i.   Find a basis for the kernel of T
   ii. Find a basis for the range of T
   iii. Verify dimension theorem

Q.10 Let $T : M_{22} \rightarrow M_{22}$ be the linear transformation defined by $T\left(\begin{bmatrix} a & b \\ c & d \end{bmatrix} \right) =$

$$\begin{bmatrix} a+b & b+c \\ a+d & b+d \end{bmatrix}$$

i.   Find a basis for $\ker(T)$
ii. Find a basis for $R(T)$

Q.11 Consider the basis $S = \{v_1, v_2, v_3\}$ for R^3, where $v_1 = (1,2,1), v_2 = (2,9,0)$, and $v_1 = (3,3,4)$, and let $T : R^3 \rightarrow R^2$ be the linear transformation for which $T(v_1) = (1,0)$, $T(v_2) = (-1,1)$, and $T(v_3) = (0,1)$. Find a formula for $T(x,y,z)$, and use that formula to find $T(7,13,7)$.

ANSWERS TO EXERCISE SET 2

1.   a. Kernel space, range
     b. $\dim(V)$
     c. Non-trivial one parameter
2.   i.   Not linear transformation (first property doesn't hold)
     ii. Not linear transformation (first property doesn't hold)
     iii. Not linear transformation (first property doesn't hold)
     iv. Yes, it is a linear transformation
     v.   Yes, it is a linear transformation
3. $T(x_1, x_2, x_3) = (-x_1 + 4x_2 - x_3, 5x_1 - 5x_2 - x_3, x_1 + 3x_3); T(2,4,-1) = (15,-9,-1)$
4. $T\begin{bmatrix} a & b \\ c & d \end{bmatrix} = a+b+c+d; T\begin{bmatrix} 1 & 3 \\ 4 & 2 \end{bmatrix} = 10$
5. $(T_2 \circ T_1)(a_0 + a_1 x) = a_0(2x+4) + a_1(2x+4)^2$
6. $(T_1 \circ T_2)\begin{bmatrix} x \\ y \end{bmatrix} = \begin{bmatrix} 2x & -y \\ 0 & 2x+2y \end{bmatrix}; (T_1 \circ T_2)\begin{bmatrix} 2 \\ 1 \end{bmatrix} = \begin{bmatrix} 4 & -1 \\ 0 & 6 \end{bmatrix}$
7. $\ker(T) = 0, R(T) = \dim(V)$
8.   (i) (a), (ii) (a)
9.   i.   Basis for $\ker(T) = \{0\}$
     ii. Basis for $R(T) = \{x, x^2, x^3\}$
     iii. $\text{rank}(T) + \text{nullity}(T) = 0 + 3 = \dim(P_2)$
10. i.   Basis for $\ker(T) = \left\{ \begin{bmatrix} 0 & 0 \\ 0 & 0 \end{bmatrix} \right\}$
     ii. Basis for $R(T) = \left\{ \begin{bmatrix} 1 & 0 \\ 1 & 0 \end{bmatrix}, \begin{bmatrix} 1 & 1 \\ 0 & 1 \end{bmatrix}, \begin{bmatrix} 0 & 1 \\ 0 & 0 \end{bmatrix}, \begin{bmatrix} 0 & 0 \\ 1 & 1 \end{bmatrix} \right\}$
11. $T(x,y,z) = (-41x + 9y + 24z, 14x - 3y - 8z); T(7,13,7) = (-2,3)$

3 Inverse Linear Transformation

In this chapter, we discuss mainly four sections. In the first section, we shall elaborate on one-one transformations and their examples. Also, we shall explain the theorem that depends upon one-one transformation using a kernel. In the second section, we define the concept of onto transformation and its examples. In the third section, we shall deliberate the isomorphism and illustrations of isomorphism between two vector spaces. In the fourth section, we shall develop an inverse linear transformation and composition of two inverse linear transformations. We shall develop the linear transformation by conveying together the two properties of one-one-ness and onto-ness, and also consider linear transformations with a bijective property. Some interesting examples are provided in Exercise Set 3.

3.1 ONE-ONE TRANSFORMATION

Definition: A linear transformation $T : V \to W$, where V and W are two vector spaces, is one-one if T maps distinct vectors in V to distance vectors in W.

3.1.1 SOME IMPORTANT RESULTS

Theorem 3.1

A linear transformation $T : V \to W$ is one-one if and only if $\text{Ker}(T) = \{0\}$.

Proof: Suppose the linear transformation has $\text{Ker}(T) = \{0\}$, then $T : V \to W$ is one-one

$\therefore T(v_1) = T(v_2)$, where $v_1, v_2 \in V$.

$$\therefore T(v_1) - T(v_2) = 0$$

$$\therefore T(v_1) + (-1)T(v_2) = 0$$

$$\therefore T(v_1 + (-1)v_2) = 0$$

$$\therefore T(v_1 - v_2) = 0.$$

Here, it is given that $\text{Ker}(T) = \{0\}$.
Therefore, $T(v_1 - v_2) = 0$ has a trivial solution.

$$v_1 - v_2 = 0 \Rightarrow v_1 = v_2.$$

Hence, a linear transformation $T : V \rightarrow W$ is one-one.

Now, if $\text{Ker}(T) = \{0\}$, then we have to show nullspace $(T) = \{0\}$.

Since T is a linear transformation, there is a 0_v in V and there is a 0_w in W. We have

$$T(0_v) = T(0_v - 0_v)$$

$$= T\left(0_v + (-1)0_v\right)$$

$$= T(0_v) + (-1)T(0_v)\left(\text{linearity property of } T\right).$$

$$= T(0_v) - T(0_v)$$

$$= 0_w.$$

Hence, $0_w \in$ nullspace (T).

On the other hand, if $v \in$ nullspace (T), then we have $T(v) = 0_w = T(0_v)$.

Since T is injective, it yields that $v = 0_v$.

Therefore, we obtain nullspace $(T) = 0_w$ and the nullity of T is zero.

T is a one-one transformation.

Theorem 3.2

A linear transformation $T : V \rightarrow W$ is one-one if and only if $\dim\left(\text{Ker}(T)\right) = 0$. From Theorem 3.1, one can easily prove Theorem 3.2.

Theorem 3.3

If A is an $n \times n$ matrix and $T_A : R^n \rightarrow R^n$ is multiplied by A, then T_A is one-one if and only if A is an invertible.

In the next section, we will discuss onto transformation and its example. Also, we will discuss the bijective transformation.

3.2   ONTO TRANSFORMATION

Let V and W be two vector spaces. A linear transformation $T : V \rightarrow W$ is onto if the range of T is W, i.e., T is onto if and only if, for every w in W, there is a v in V such that $T(V) = W$. An onto transformation is also called surjective transformation.

Theorem 3.4

A linear transformation $T : V \rightarrow W$ is onto if and only if $\text{rank}(T) = \dim(W)$.

Theorem 3.5

Let $T : V \to W$ be a linear transformation, and let $\text{rank}(T) = \dim(W)$. (i) If T is one-one, then it is onto. (ii) If T is onto, then it is one-one.

Definition: Bijective Transformation

If a transformation $T : V \to W$ is both one-to-one and onto, then it is called bijective transformation.

Example 3.1

Determine whether the linear transformation $T : R^2 \to R^2$, where $T(x,y) = (x+y, x-y)$, is one-one, onto, or both or neither.

Solution

A linear transformation is one-one if and only if $\ker(T) = \{0\}$
   Let

$$T(x,y) = \bar{0}$$

$$\therefore (x+y, x-y) = (0,0)$$

$$x+y = 0, x-y = 0.$$

Solving these equations,

$$x = 0, \quad y = 0.$$

The solution is trivial; hence, $\ker(T) = \{0\}$.
   Hence, T is one-one.
   A linear transformation is onto if $R(T) = W$.
   Let $v = (x,y)$ and $w = (a,b)$ be in R^2, where a and b are the real numbers such that $T(V) = W$.

$$T(x,y) = (a,b)$$

$$\because (x+y, x-y) = (a,b)$$

$$\because x+y = a, x-y = b.$$

Solving these equations,

$$x = \frac{a+b}{2}, \quad y = \frac{a-b}{2}.$$

Thus, for every $w = (a,b)$ in R^2, there exists a $v = \left(\frac{a+b}{2}, \frac{a-b}{2} \right)$ in R^2.
   Hence, T is onto.

Example 3.2

Determine whether the linear transformation $T: R^2 \rightarrow R^3$, where $T(x,y) = (x-y, y-x, 2x-2y)$, is one-one, onto, or both or neither.

Solution

A linear transformation is one-one if and only if $\ker(T) = \{0\}$.
  Let

$$T(x,y) = \bar{0}$$

$$\therefore (x-y, y-x, 2x-2y) = (0,0,0)$$

$$x-y = 0, \quad y-x = 0, \quad 2x-2y = 0.$$

Solving these equations, $x = y$

$$x = t, \quad y = t, \text{ where, } t \in R.$$

Solution set is $S = \{(x,y) = t(1,1) / t \in R\}$.
  The solution is trivial; hence, $\ker(T) = \{(x,y) = t(1,1)/t \in R\} \neq \{0\}$.
  Hence, T is not one-one.
  A linear transformation is onto if $R(T) = W$.
  Let $v = (x,y)$ in R^2 and $w = (a,b,c) \in R^3$, where a, b, and c are the real numbers such that $T(V) = W$

$$T(x,y) = (a,b,c)$$

$$\because (x-y, y-x, 2x-2y) = (a,b,c)$$

$$\because x-y = a, \quad y-x = b, \quad 2x-2y = c.$$

Solving these equations,
  the system has a solution only when $a = -b = \dfrac{c}{2}$ not for all a, b, and c.
  Hence, T is not onto.

Example 3.3

Determine whether the linear transformation $T: P_2 \rightarrow R^3$, where

$$T(a+bx+cx^2) = \begin{bmatrix} 2a-b \\ a+b-3c \\ c-a \end{bmatrix}, \text{ is one-one, onto, or both or neither.}$$

Solution

Let $T(a+bx+cx^2) = 0$

$$\therefore \begin{bmatrix} 2a-b \\ a+b-3c \\ c-a \end{bmatrix} = \begin{bmatrix} 0 \\ 0 \\ 0 \end{bmatrix}.$$

Comparing both sides,

$$\therefore 2a-b=0, \quad a+b-3c=0, \quad c-a=0.$$

Solving the above equations by Gauss elimination, we get

$$\text{Augmented matrix} = \begin{bmatrix} 2 & -1 & 0 & | & 0 \\ 1 & 1 & -3 & | & 0 \\ -1 & 0 & 1 & | & 0 \end{bmatrix}.$$

Take $R_2 \leftrightarrow R_1$

$$\approx \begin{bmatrix} 1 & 1 & -3 & | & 0 \\ 2 & -1 & 0 & | & 0 \\ -1 & 0 & 1 & | & 0 \end{bmatrix}.$$

By taking $R_2 \rightarrow R_2 - 2R_1, \ R_3 \rightarrow R_3 + R_1$

$$\approx \begin{bmatrix} 1 & 1 & -3 & | & 0 \\ 0 & -3 & 6 & | & 0 \\ 0 & 1 & -2 & | & 0 \end{bmatrix}.$$

$$\frac{-1}{3} R_2$$

$$\approx \begin{bmatrix} 1 & 1 & -3 & | & 0 \\ 0 & 1 & -2 & | & 0 \\ 0 & 1 & -2 & | & 0 \end{bmatrix}.$$

$$R_3 \rightarrow R_3 - R_2$$

$$\approx \begin{bmatrix} 1 & 1 & -3 & | & 0 \\ 0 & 1 & -2 & | & 0 \\ 0 & 0 & 0 & | & 0 \end{bmatrix}.$$

By taking substitution

$$a+b-3c = 0, \quad b-2c = 0, \quad c = t \in R$$

$$a = t, b = 2t, c = t \in R.$$

Solution set is $S = \{(a,b,c) = t(1,2,1)/t \in R\}$.

The solution is trivial; hence, $\ker(T) = \{(a,b,c) = t(1,2,1)/t \in R\} \neq \{0\}$.
Hence, T is not one-one.
In the next section, we will discuss isomorphism and its example.

3.3 ISOMORPHISM

A bijective transformation from V and W is known as an isomorphism between V and W. It is denoted by $V \cong W$.

Theorem 3.6

Every real n-dimensional vector space is isomorphic to R^n.
Proof: Let V be a real n-dimensional vector space.
Therefore, we have to prove a linear transformation $T : V \to R^n$ is one-one and onto.
Let $S = \{v_1, v_2, ..., v_n\}$ be any basis for V.
And let $u = k_1 v_1 + k_2 v_2 + \cdots + k_n v_n$ be the representation of a vector u in V as a linear combination of the basis vectors, and let $T : V \to R^n$ be the coordinate map $T(u) = (u)_s = (k_1, k_2, ..., k_n)$.
For proving isomorphism, first to prove linearity.
Let u and v be vectors in V, let c be a scalar, then
$u = k_1 v_1 + k_2 v_2 + \cdots + k_n v_n$ and $v = d_1 v_1 + d_2 v_2 + \cdots + d_n v_n$ are the representations of u and v as a linear combination of the basis vectors. Then, it follows that

$$T(cu) = T(ck_1 v_1 + ck_2 v_2 + \cdots + ck_n v_n)$$

$$= (ck_1, ck_2, ..., ck_n)$$

$$= c(k_1, k_2, ..., k_n)$$

$$= cT(u)$$

and

$$T(u+v) = T((k_1 + d_1)v_1 + (k_2 + d_2)v_2 + \cdots + (k_n + d_n)v_n)$$

$$= (k_1 + d_1, k_2 + d_2, ..., k_n + d_n)$$

$$= (k_1, k_2, ..., k_n) + (d_1, d_2, ..., d_n)$$

$$= T(u) + T(v),$$

which shows that T is linear. For one-one transformation, we must show that if u and v are distinct vectors in V, then that are their images in R^n. But if $u \neq v$ and if the representations of these vectors in terms of the basis vector areas in $u = k_1 v_1 + k_2 v_2 + \cdots + k_n v_n$ and $v = d_1 v_1 + d_2 v_2 + \cdots + d_n v_n$, then we must have $k_i \neq d_i$ for at least one i. Thus,

$$T(u) = (k_1, k_2, \ldots, k_n) \neq (d_1, d_2, \ldots, d_n) = T(v),$$

which shows that u and v have distinct images under T. Finally, the transformation T is $w = (k_1, k_2, \ldots, k_n)$, which is any vector in R^n, then it follows from $T(u) = (u)_s = (k_1, k_2, \ldots, k_n)$ that w is the image under T of the vector $u = k_1 v_1 + k_2 v_2 + \cdots + k_n v_n$.

Therefore, every real n-dimensional vector space is isomorphic to R^n.

Example 3.4

Show that R^4 and P_3 are isomorphic.

Solution

Consider the linear transformation $T : R^4 \to P_3$, which is defined by

$$T(a,b,c,d) = ax^3 + bx^2 + cx + d.$$

Obviously, $\dim(R^4) = \dim(P_3) = 4$.

Let, $T(a,b,c,d) = \bar{0}$

$$\Rightarrow ax^3 + bx^2 + cx + d = \bar{0}$$

$$\Rightarrow a = 0,\ b = 0,\ c = 0,\ d = 0$$

$$\therefore \ker(T) = \{0\}.$$

$\therefore T$ is one-one and onto.

$$\therefore R^4 \cong P_3.$$

R^4 and P_3 are isomorphic.

Example 3.5

Show that P_3 and M_{22} are isomorphic.

Solution

Consider the linear transformation $T : P_3 \to M_{22}$, which is defined by

$$T\left(ax^3 + bx^2 + cx + d\right) = \begin{bmatrix} a & b \\ c & d \end{bmatrix}.$$

Obviously, $\dim(P_3) = \dim(M_{22}) = 4$.

Let $T\left(ax^3 + bx^2 + cx + d\right) = \bar{0}$

$$\Rightarrow \begin{bmatrix} a & b \\ c & d \end{bmatrix} = \bar{0}$$

$$\Rightarrow a = 0, \ b = 0, \ c = 0, \ d = 0$$

$$\therefore \ker(T) = \{0\}.$$

$\therefore T$ is one-one and onto.

$$\therefore P_3 \cong M_{22}.$$

P_3 and M_{22} are isomorphic.

In the next section, we will discuss the inverse linear transformation and its composition.

3.4   INVERSE LINEAR TRANSFORMATION

If $T : V \to W$ is a linear transformation, then the range of T is the subspace of W consisting of all images of vectors in V under T. If T is one-to-one, then each vector w in $R(T)$ is the image of a unique vector u in V. Hence, inverse linear transformation $T^{-1} : W \to V$ maps w back into v.

Theorem 3.7

If $T_1 : U \to V$ and $T_2 : V \to W$ are one-one transformations, then (i) $T_2 \circ T_1$ is one-one and (ii) $(T_2 \circ T_1)^{-1} = T_1^{-1} \circ T_2^{-1}$.

Example 3.6

Let $T : R^3 \to R^3$ be the linear operator defined by the formula $T(x_1, x_2, x_3) = (x_1 - x_2 + x_3, 2x_2 - x_3, 2x_1 + 3x_2)$. Determine whether T is one-one. If so, find $T^{-1}(x_1, x_2, x_3)$.

Solution

The standard matrix of T is $T = \begin{bmatrix} 1 & -1 & 1 \\ 0 & 2 & -1 \\ 2 & 3 & 0 \end{bmatrix}$

$$\det(A) = 1(0+3) + 1(0+2) + 1(0-4)$$

$$= 3 + 2 - 4$$

$$= 1 \neq 0.$$

Therefore, T is one-one.

Using the Gauss-Jordan method to find T^{-1}

$$[T \mid I] = \begin{bmatrix} 1 & -1 & 1 & 1 & 0 & 0 \\ 0 & 2 & -1 & 0 & 1 & 0 \\ 2 & 3 & 0 & 0 & 0 & 1 \end{bmatrix}$$

$$R_3 \to R_3 - 2R_1$$

$$\approx \begin{bmatrix} 1 & -1 & 1 & 1 & 0 & 0 \\ 0 & 2 & -1 & 0 & 1 & 0 \\ 0 & 5 & -2 & -2 & 0 & 1 \end{bmatrix}$$

$$R_2 \to \frac{1}{2} R_2$$

$$\approx \begin{bmatrix} 1 & -1 & 1 & 1 & 0 & 0 \\ 0 & 1 & -\dfrac{1}{2} & 0 & \dfrac{1}{2} & 0 \\ 0 & 5 & -2 & -2 & 0 & 1 \end{bmatrix}$$

$$R_3 \to R_3 - 5R_2, \; R_1 \to R_1 + R_2$$

$$\approx \begin{bmatrix} 1 & 0 & \dfrac{1}{2} & 1 & \dfrac{1}{2} & 0 \\ 0 & 1 & -\dfrac{1}{2} & 0 & \dfrac{1}{2} & 0 \\ 0 & 0 & \dfrac{1}{2} & -2 & -\dfrac{5}{2} & 1 \end{bmatrix}$$

$$R_3 \to 2R_3$$

$$\approx \begin{bmatrix} 1 & 0 & \dfrac{1}{2} & 1 & \dfrac{1}{2} & 0 \\ 0 & 1 & -\dfrac{1}{2} & 0 & \dfrac{1}{2} & 0 \\ 0 & 0 & 1 & -4 & -5 & 2 \end{bmatrix}$$

$$R_2 \rightarrow R_2 + \frac{1}{2}R_3, \ R_1 \rightarrow R_1 - \frac{1}{2}R_3$$

$$\left[I \mid T^{-1} \right] = \begin{bmatrix} 1 & 0 & 0 & 3 & 3 & -1 \\ 0 & 1 & 0 & -2 & -2 & 1 \\ 0 & 0 & 1 & -4 & -5 & 2 \end{bmatrix}$$

$$T^{-1} = \begin{bmatrix} 3 & 3 & -1 \\ -2 & -2 & 1 \\ -4 & -5 & 2 \end{bmatrix}$$

$$T^{-1}\left(\begin{bmatrix} x_1 \\ x_2 \\ x_3 \end{bmatrix} \right) = \begin{bmatrix} 3 & 3 & -1 \\ -2 & -2 & 1 \\ -4 & -5 & 2 \end{bmatrix} \begin{bmatrix} x_1 \\ x_2 \\ x_3 \end{bmatrix} = \begin{bmatrix} 3x_1 + 3x_2 - x_3 \\ -2x_1 - 2x_2 + x_3 \\ -4x_1 - 5x_2 + 2x_3 \end{bmatrix}$$

$$T^{-1}(x_1,x_2,x_3) = \left(3x_1 + 3x_2 - x_3, \ -2x_1 - 2x_2 + x_3, \ -4x_1 - 5x_2 + 2x_3 \right).$$

Example 3.7

Consider $T : R^2 \rightarrow R^2$ be the linear operator given by the formula $T(x,y) = (x + ky, -y)$. Show that T is one-one for every real value of k and that $T^{-1} = T$.

Solution

Here, the liner operator $T(x,y) = (x + ky, -y)$ is given.

The standard matrix for this operator is $T = \begin{bmatrix} 1 & k \\ 0 & -1 \end{bmatrix}$.

By Theorem 3.1, $Ker(T) = \{0\}$, because $|T| = -1 \neq 0$ for any value of k. $T(X) = 0$ has a unique solution.

Therefore, T is the one-one transformation for any value of k.

$$T^{-1} = \frac{adjT}{|T|} = \frac{1}{-1} \begin{bmatrix} -1 & -k \\ 0 & 1 \end{bmatrix}$$

$$= \begin{bmatrix} 1 & k \\ 0 & -1 \end{bmatrix}$$

$$= T.$$

Hence, the proof.

Example 3.8

Determine $T^{-1}(x_1,x_2,x_3)$ if $T : R^3 \rightarrow R^3$ is the linear operator defined by the formula $T(x_1,x_2,x_3) = (3x_1 + x_2, -2x_1 - 4x_2 + 3x_3, 5x_1 + 4x_2 - 2x_3)$.

Solution

Let the standard matrix of the transformation
$$T(x_1,x_2,x_3) = (3x_1 + x_2, -2x_1 - 4x_2 + 3x_3, 5x_1 + 4x_2 - 2x_3) \text{ is}$$

$$T = \begin{bmatrix} 3 & 1 & 0 \\ -2 & -4 & 3 \\ 5 & 4 & -2 \end{bmatrix}$$

$$\det(T) = 3(-4) - 1(-4) - 0(12)$$
$$= -12 + 4$$
$$= -8 \neq 0.$$

Therefore, T is one-one transformation.
Use the Gauss-Jordan method for finding T^{-1}

$$[T \mid I] = \begin{bmatrix} 3 & 1 & 0 & | & 1 & 0 & 0 \\ -2 & -4 & 3 & | & 0 & 1 & 0 \\ 5 & 4 & -2 & | & 0 & 0 & 1 \end{bmatrix}$$

$$R_1 \to \frac{1}{3} R_1$$

$$\approx \begin{bmatrix} 1 & \frac{1}{3} & 0 & | & \frac{1}{3} & 0 & 0 \\ -2 & -4 & 3 & | & 0 & 1 & 0 \\ 5 & 4 & -2 & | & 0 & 0 & 1 \end{bmatrix}$$

$$R_2 \to R_2 + 2R_1, R_3 \to R_3 - 5R_1$$

$$\approx \begin{bmatrix} 1 & \frac{1}{3} & 0 & | & \frac{1}{3} & 0 & 0 \\ 0 & \frac{-10}{3} & 3 & | & \frac{2}{3} & 1 & 0 \\ 0 & \frac{7}{3} & -2 & | & \frac{-5}{3} & 0 & 1 \end{bmatrix}$$

$$R_2 \to -\frac{3}{10} R_2$$

$$\approx \begin{bmatrix} 1 & \frac{1}{3} & 0 & | & \frac{1}{3} & 0 & 0 \\ 0 & 1 & -\frac{9}{10} & | & -\frac{1}{5} & -\frac{3}{10} & 0 \\ 0 & \frac{7}{3} & -2 & | & \frac{-5}{3} & 0 & 1 \end{bmatrix}$$

$$R_1 \to R_1 - \frac{1}{3} R_2, R_3 \to R_3 - \frac{7}{3} R_2$$

$$\approx \begin{bmatrix} 1 & 0 & \dfrac{3}{10} & \bigg| & \dfrac{2}{5} & \dfrac{1}{10} & 0 \\ 0 & 1 & \dfrac{-9}{10} & \bigg| & \dfrac{-1}{5} & \dfrac{-3}{10} & 0 \\ 0 & 0 & \dfrac{1}{10} & \bigg| & \dfrac{-6}{5} & \dfrac{7}{10} & 1 \end{bmatrix}$$

$$R_3 \to 10R_3$$

$$\approx \begin{bmatrix} 1 & 0 & \dfrac{3}{10} & \bigg| & \dfrac{2}{5} & \dfrac{1}{10} & 0 \\ 0 & 1 & \dfrac{-9}{10} & \bigg| & \dfrac{-1}{5} & \dfrac{-3}{10} & 0 \\ 0 & 0 & 1 & \bigg| & -12 & 7 & 10 \end{bmatrix}$$

$$R_1 \to R_1 - \dfrac{3}{10}R_3, R_2 \to R_2 + \dfrac{9}{10}R_3$$

$$\approx \begin{bmatrix} 1 & 0 & 0 & \bigg| & 4 & -2 & -3 \\ 0 & 1 & 0 & \bigg| & -11 & 6 & 9 \\ 0 & 0 & 1 & \bigg| & -12 & 7 & 10 \end{bmatrix}$$

$$\left[T^{-1} \right] = \begin{bmatrix} 4 & -2 & -3 \\ -11 & 6 & 9 \\ -12 & 7 & 10 \end{bmatrix}.$$

Then, the inverse transformation is

$$T^{-1}\left(\begin{bmatrix} x_1 \\ x_2 \\ x_3 \end{bmatrix} \right) = \begin{bmatrix} 4 & -2 & -3 \\ -11 & 6 & 9 \\ -12 & 7 & 10 \end{bmatrix} \begin{bmatrix} x_1 \\ x_2 \\ x_3 \end{bmatrix} = \begin{bmatrix} 4x_1 - 2x_2 - 3x_3 \\ -11x_1 + 6x_2 + 9x_3 \\ -12x_1 + 7x_2 + 10x_3 \end{bmatrix}$$

$$T^{-1}(x_1,x_2,x_3) = (4x_1 - 2x_2 - 3x_3, -11x_1 + 6x_2 + 9x_3, -12x_1 + 7x_2 + 10x_3).$$

Example 3.9

Let $T: R^2 \to R^2$ and $S: R^2 \to R^2$ be the linear operators given by the formula $T(x,y) = (x+y, x-y)$ and $S(x,y) = (2x+y, x-2y)$.

    i. Show that T and S are one-one.
    ii. Find formulas for $T^{-1}(x,y)$ and $S^{-1}(x,y)$ and $(T \circ S)^{-1}(x,y)$.
    iii. Verify $(T \circ S)^{-1} = S^{-1} \circ T^{-1}$.

Solution

i. T and S are one-one if $\ker(T) = \{0\}$ and $\ker(S) = \{0\}$.
   For that, $T(X) = 0$

$$T(x,y) = \bar{0}$$
$$\therefore (x+y, x-y) = (0,0) \cdot$$
$$x+y = 0, \ x-y = 0$$

Solving these equations,

$$x = 0, \ y = 0 \cdot$$

The solution is trivial; hence, $\ker(T) = \{0\}$.
   Hence, T is one-one.
   Similarly, $S(X) = 0$

$$S(x,y) = \bar{0}$$
$$\therefore (2x+y, x-2y) = (0,0) \cdot$$
$$2x+y = 0, \ x-2y = 0$$

Solving these equations,

$$x = 0, \ y = 0 \cdot$$

The solution is trivial; hence, $\ker(S) = \{0\}$.
   Hence, S is one-one.

ii. Standard matrix for $T = \begin{bmatrix} 1 & 1 \\ 1 & -1 \end{bmatrix}$

$$\because T^{-1} = \frac{\mathrm{adj}(T)}{|T|} = \frac{1}{-2} \begin{bmatrix} -1 & -1 \\ -1 & 1 \end{bmatrix} = \begin{bmatrix} \frac{1}{2} & \frac{1}{2} \\ \frac{1}{2} & -\frac{1}{2} \end{bmatrix}$$

$$\because T^{-1}(x,y) = \left(\frac{x+y}{2}, \frac{x-y}{2} \right) \cdot$$

Standard matrix for $S = \begin{bmatrix} 2 & 1 \\ 1 & -2 \end{bmatrix}$

$$\because S^{-1} = \frac{\mathrm{adj}(S)}{|S|} = \frac{-1}{5} \begin{bmatrix} -2 & -1 \\ -1 & 2 \end{bmatrix} = \begin{bmatrix} \frac{2}{5} & \frac{1}{5} \\ \frac{1}{5} & \frac{-2}{5} \end{bmatrix}$$

$$\because T^{-1}(x,y) = \left(\frac{2x+y}{5}, \frac{x-2y}{5} \right)$$

$$T \circ S = T \cdot S = \begin{bmatrix} 1 & 1 \\ 1 & -1 \end{bmatrix} \cdot \begin{bmatrix} 2 & 1 \\ 1 & -2 \end{bmatrix}$$

$$= \begin{bmatrix} 3 & -1 \\ 1 & 3 \end{bmatrix}$$

$$(T \cdot S)^{-1} = \frac{T \cdot S}{|T \cdot S|} = \frac{1}{10} \begin{bmatrix} 3 & 1 \\ -1 & 3 \end{bmatrix} = \begin{bmatrix} \dfrac{3}{10} & \dfrac{1}{10} \\ -\dfrac{1}{10} & \dfrac{3}{10} \end{bmatrix}$$

$$(T \circ S)^{-1}(x,y) = \begin{bmatrix} \dfrac{3}{10} & \dfrac{1}{10} \\ -\dfrac{1}{10} & \dfrac{3}{10} \end{bmatrix} \begin{bmatrix} x \\ y \end{bmatrix}$$

$$(T \circ S)^{-1}(x,y) = \left(\frac{3x+y}{10}, \frac{-x+3y}{10} \right).$$

iii. From (ii) LHS of $(T \circ S)^{-1} = S^{-1} \circ T^{-1}$ is

$$(T \circ S)^{-1} = \begin{bmatrix} \dfrac{3}{10} & \dfrac{1}{10} \\ -\dfrac{1}{10} & \dfrac{3}{10} \end{bmatrix}.$$

From (ii) $T^{-1} = \begin{bmatrix} \dfrac{1}{2} & \dfrac{1}{2} \\ \dfrac{1}{2} & -\dfrac{1}{2} \end{bmatrix}$ and $S^{-1} = \begin{bmatrix} \dfrac{2}{5} & \dfrac{1}{5} \\ \dfrac{1}{5} & \dfrac{-2}{5} \end{bmatrix}$

$$S^{-1} \circ T^{-1} = S^{-1} \cdot T^{-1} = \begin{bmatrix} \dfrac{2}{5} & \dfrac{1}{5} \\ \dfrac{1}{5} & \dfrac{-2}{5} \end{bmatrix} \begin{bmatrix} \dfrac{1}{2} & \dfrac{1}{2} \\ \dfrac{1}{2} & -\dfrac{1}{2} \end{bmatrix}$$

$$= \begin{bmatrix} \dfrac{3}{10} & \dfrac{1}{10} \\ -\dfrac{1}{10} & \dfrac{3}{10} \end{bmatrix}$$

$$\therefore (T \circ S)^{-1} = S^{-1} \circ T^{-1}.$$

Hence, the proof.

EXERCISE SET 3

Q.1   Determine whether the linear transformation is one-one by using kernel
       a. $T : R^2 \to R^2$, where $T(x,y) = (2y, 3x)$
       b. $T : R^2 \to R^3$, where $T(x,y) = (2x, 5y, x+y)$
       c. $T : R^3 \to R^2$, where $T(x,y,z) = (x + 2y + 3z, x - 5y - 8z)$
       d. $T : R^2 \to R^3$, where $T(x,y) = (x - y, y - x, 2x - 2y)$
       e. $T : R^2 \to R^2$, where $T(x,y) = (0, 2x + 3y)$

Q.2   Which of the following transformations is bijective?
       a. $T : P_2 \to P_3$, where $T(P(x)) = xP(x)$
       b. $T : M_{22} \to M_{22}$, where $T(A) = A^T$
       c. $T : P_3 \to R^3$, where $T(a + bx + cx^2 + dx^3) = (b, c, d)$
       d. $T : R^4 \to R^3$, where $T(x,y,z,w) = (x,y,0)$

Q.3   Let $T_1 : P_2 \to P_3$ and $T_2 : P_3 \to P_3$ be the linear transformations given by the
       formulas
       $T_1(P(x)) = xP(x)$ and $T_2(P(x)) = P(x+1)$
       a. Find formulas for $T_1^{-1}(P(x))$, $T_2^{-1}(P(x))$ and $(T_2 \circ T_1)^{-1}(P(x))$
       b. Verify that $(T_2 \circ T_1)^{-1} = T_1^{-1} \circ T_2^{-1}$

Q.4   Let $T : P_1 \to R^2$ be the function defined by $T(p(x)) = (p(0), p(1))$
       i. Find $T(1 - 2x)$
       ii. Show that T is one-to-one
       iii. Find $T^{-1}(2,3)$

Q.5   Check whether the following are isomorphism or not?
       a. $c_0 + c_1 x \to (c_0 - c_1, c_1)$ from P_1 to R^2
       b. $(a,b,c,d) \to a + bx + cx^2 + (d+1)x^3$ from R^4 to P_3
       c. $A \to A^T$ from M_{nn} to M_{nn}

Q.6   Let $T : R^3 \to R^3$ be a multiplier of A, find $T^{-1}\left(\begin{bmatrix} x_1 \\ x_2 \\ x_3 \end{bmatrix}\right)$

(i) $A = \begin{bmatrix} 1 & 0 & 1 \\ 0 & 1 & 1 \\ 1 & 1 & 0 \end{bmatrix}$, (ii) $A = \begin{bmatrix} 1 & -1 & 1 \\ 0 & 2 & -1 \\ 2 & 3 & 0 \end{bmatrix}$

Q.7   Write whether the following statements are true or false.
       a. The vector space R^2 and P_2 are isomorphic.
       b. If the kernel of a linear transformation $T : P_3 \rightarrow P_3$ is $\{0\}$, then T is an
          isomorphism.
       c. Every linear transformation from M_{33} to P_9 is an isomorphism.
       d. There is a subspace of M_{23} that is isomorphism to R^4.
       e. Isomorphic finite-dimensional vector spaces must have the same
          number of basis vectors.
       f. R^n is isomorphic to a subspace of R^{n+1}.

ANSWERS TO EXERCISE SET 3

Q.1   a. One-to-one
         b. One-to-one
         c. Not one-to-one
         d. Not one-to-one
         e. Not one-to-one
Q.2   a. No (not onto)
         b. Yes
         c. No (not one-to-one)
         d. No (not one-to-one)

Q.3 $T_1^{-1}(p(x)) = \dfrac{p(x)}{x}, T_2^{-1}(p(x)) = p(x-1), (T_2 \circ T_1)^{-1}(p(x)) = \dfrac{p(x-1)}{x}$

Q.4   i. $(1,-1)$
         ii. $(2+x)$
Q.5   a. Isomorphism
         b. Not isomorphism
         c. Isomorphism

Q.6   i. $T^{-1}(x_1,x_2,x_3) = \left(\dfrac{1}{2}x_1 - \dfrac{1}{2}x_2 + \dfrac{1}{2}x_3, \ -\dfrac{1}{2}x_1 + \dfrac{1}{2}x_2 + \dfrac{1}{2}x_3, \ \dfrac{1}{2}x_1 + \dfrac{1}{2}x_2 - \dfrac{1}{2}x_3 \right)$
        ii. $T^{-1}(x_1,x_2,x_3) = \left(3x_1 + 3x_2 - x_3, \ -2x_1 - 2x_2 + x_3, \ -4x_1 - 5x_2 + 2x_3 \right)$

Q.7   a. False
         b. True
         c. False
         d. True
         e. True

4 Matrices of General Linear Transformations

In this chapter, we shall develop mainly two sections. In the first section, we shall study the concept of a matrix of general transformation. We shall find a matrix for the linear transformation concerning the bases for vector spaces in the first section. We shall also solve the examples of the matrix for the linear transformation and linear operator. In the second section, we shall discuss the similarity of the two matrices as the similarity is a very important concept to convert complicated matrices in the form of matrix as simple as possible (e.g., diagonal or triangular matrix). Also, we shall explain the properties of similarities of two matrices together with some examples. Lastly, we will provide some examples in the exercise with answers, which will help learners to practise the concepts.

4.1 MATRIX OF GENERAL TRANSFORMATIONS

Let V be an n-dimensional vector space and W be an m-dimensional vector space. If we choose bases S_1 and S_2 for V and W, respectively, then for each x in V, the coordinate vector $[X]_{S_1}$ will be a vector in R^n and the coordinate vector $[T(X)]_{S_2}$ will be a vector in R^m.

Suppose $T : V \rightarrow W$ is a linear transformation. If we assume A to be the standard matrix for this transformation, then $A[X]_{S_1} = [T(X)]_{S_2}$.

The above matrix A is called the matrix for linear transformation T with respect to the ordered bases S_1 and S_2.

First, we show how we can compute a matrix.

Let $S_1 = \{u_1, u_2, \ldots, u_n\}$ and $S_2 = \{v_1, v_2, \ldots, v_m\}$ be the ordered bases for the n-dimensional space V and m-dimensional space W, respectively.

Let A denote the $m \times n$ matrix. Thus,

$$A = \begin{bmatrix} a_{11} & a_{12} & . & . & . & a_{1n} \\ a_{21} & a_{22} & . & . & . & a_{2n} \\ . & . & . & . & . & . \\ . & . & . & . & . & . \\ . & . & . & . & . & . \\ a_{m1} & a_{m2} & . & . & . & a_{mn} \end{bmatrix}$$

such that $A[X]_{S_1} = [T(X)]_{S_2}$ holds for all vectors in V, i.e., A times the coordinate vector X equals the coordinate vector of the image $T(X)$ of X. That is, equation $A[X]_{S_1} = [T(X)]_{S_2}$ holds for the basis vectors $u_1, u_2, \ldots, u_n$; i.e.,

$$A[u_1]_{S_1} = [T(u_1)]_{S_2}, A[u_2]_{S_1} = [T(u_2)]_{S_2}, \ldots, A[u_n]_{S_1} = [T(u_n)]_{S_2}. \qquad (4.1)$$

But

$$[u_1]_{S_1} = \begin{bmatrix} 1 \\ 0 \\ 0 \\ \cdot \\ \cdot \\ 0 \end{bmatrix}, [u_2]_{S_1} = \begin{bmatrix} 0 \\ 1 \\ 0 \\ \cdot \\ \cdot \\ 0 \end{bmatrix}, \ldots [u_n]_{S_1} = \begin{bmatrix} 0 \\ 0 \\ 0 \\ \cdot \\ \cdot \\ 1 \end{bmatrix}.$$

$$\therefore A[u_1]_{S_1} = \begin{bmatrix} a_{11} \\ a_{21} \\ \cdot \\ \cdot \\ \cdot \\ a_{m1} \end{bmatrix}, \therefore A[u_2]_{S_1} = \begin{bmatrix} a_{12} \\ a_{22} \\ \cdot \\ \cdot \\ \cdot \\ a_{m2} \end{bmatrix}, \ldots, \therefore A[u_n]_{S_1} = \begin{bmatrix} a_{1n} \\ a_{2n} \\ \cdot \\ \cdot \\ \cdot \\ a_{mn} \end{bmatrix}.$$

Substituting these results into equation (4.1), we get

$$\begin{bmatrix} a_{11} \\ a_{21} \\ \cdot \\ \cdot \\ \cdot \\ a_{m1} \end{bmatrix} = [T(u_1)]_{S_2}, \begin{bmatrix} a_{12} \\ a_{22} \\ \cdot \\ \cdot \\ \cdot \\ a_{m2} \end{bmatrix} = [T(u_2)]_{S_2}, \ldots, \begin{bmatrix} a_{1n} \\ a_{2n} \\ \cdot \\ \cdot \\ \cdot \\ a_{mn} \end{bmatrix} = [T(u_n)]_{S_2},$$

which shows that the successive columns of A are the coordinate vectors of $T(u_1), T(u_2), \ldots, T(u_n)$ with respect to the basis S_2.

Thus, the matrix for linear transformation T with respect to the basis S_1 and S_2 is

$$A = \left[[T(u_1)]_{S_2} \,\middle|\, [T(u_2)]_{S_2} , \ldots, [T(u_n)]_{S_2} \right].$$

Thus A can be denoted by $[T]_{S_1, S_2}$.

Therefore, $[T]_{S_1, S_2} = \left[[T(u_1)]_{S_2} \,\middle|\, [T(u_2)]_{S_2} , \ldots, [T(u_n)]_{S_2} \right]$.

Also, this matrix has the property $[T]_{S_1, S_2}[X]_{S_1} = [T(X)]_{S_2}$.

When $V = W$ (i.e. $T : V \to V$ is a linear operator), it is usual to take $S_1 = S_2 = S$ when constructing a matrix for T. The resulting matrix, in this case, is called a matrix for T with respect to the basis S and is denoted by $[T]_S$ instead of $[T]_{S_1, S_2}$. If $S = \{u_1, u_2, \ldots, u_n\}$, then the formula for $[T]_{S_1, S_2}$ becomes $[T]_S = \left[[T(u_1)]_S \,\middle|\, [T(u_2)]_S , \ldots, [T(u_n)]_S \right]$.

Also, this matrix has the property $[T]_S[X]_S = [T(X)]_S$.

Example 4.1

Let $T: R^2 \rightarrow R^3$ be the linear transformation defined by $T\left(\begin{bmatrix} x_1 \\ x_2 \end{bmatrix}\right) =$

$$\begin{bmatrix} x_2 \\ -5x_1 + 13x_2 \\ -7x_1 + 16x_2 \end{bmatrix}.$$

Find the matrix of the transformation T with respect to the bases $S_1 = \{v_1, v_2\}$

for R^2 and $S_2 = \{w_1, w_2, w_3\}$ for R^3, where $v_1 = \begin{bmatrix} 3 \\ 1 \end{bmatrix}$, $v_1 = \begin{bmatrix} 5 \\ 2 \end{bmatrix}$, $w_1 = \begin{bmatrix} 1 \\ 0 \\ -1 \end{bmatrix}$,

$w_2 = \begin{bmatrix} -1 \\ 2 \\ 2 \end{bmatrix}$, $w_3 = \begin{bmatrix} 0 \\ 1 \\ 2 \end{bmatrix}$.

Solution

Here, $T(v_1) = T\left(\begin{bmatrix} 3 \\ 1 \end{bmatrix}\right) = \begin{bmatrix} 1 \\ -5(3)+13(1) \\ -7(3)+16(1) \end{bmatrix} = \begin{bmatrix} 1 \\ -2 \\ -5 \end{bmatrix}$

$T(v_2) = T\left(\begin{bmatrix} 5 \\ 2 \end{bmatrix}\right) = \begin{bmatrix} 2 \\ -5(5)+13(2) \\ -7(5)+16(2) \end{bmatrix} = \begin{bmatrix} 2 \\ 1 \\ -3 \end{bmatrix}$.

Let us express v_1 as a linear combination of w_1, w_2, w_3

$$T(v_1) = k_1 w_1 + k_2 w_2 + k_3 w_3.$$

$$\therefore \begin{bmatrix} 1 \\ -2 \\ -5 \end{bmatrix} = k_1 \begin{bmatrix} 1 \\ 0 \\ -1 \end{bmatrix} + k_2 \begin{bmatrix} -1 \\ 2 \\ 2 \end{bmatrix} + k_3 \begin{bmatrix} 0 \\ 1 \\ 2 \end{bmatrix}.$$

$$\therefore \begin{bmatrix} 1 \\ -2 \\ -5 \end{bmatrix} = \begin{bmatrix} k_1 - k_2 \\ 2k_2 + k_3 \\ -k_1 + 2k_2 + 2k_3 \end{bmatrix}.$$

Comparing both sides,

$$k_1 - k_2 = 1, 2k_2 + k_3 = -2, -k_1 + 2k_2 + 2k_3 = -5.$$

Using Gauss elimination method,

$$k_1 = 1, k_2 = 0, k_3 = -2$$

$$T(v_1) = w_1 - 2w_3$$

$$\left[T(v_1)\right]_{s_2} = \begin{bmatrix} 1 \\ 0 \\ -2 \end{bmatrix}.$$

Express v_2 as a linear combination of w_1, w_2, w_3

$$T(v_2) = k_1 w_1 + k_2 w_2 + k_3 w_3.$$

$$\therefore \begin{bmatrix} 2 \\ 1 \\ -3 \end{bmatrix} = k_1 \begin{bmatrix} 1 \\ 0 \\ -1 \end{bmatrix} + k_2 \begin{bmatrix} -1 \\ 2 \\ 2 \end{bmatrix} + k_3 \begin{bmatrix} 0 \\ 1 \\ 2 \end{bmatrix}.$$

$$\therefore \begin{bmatrix} 2 \\ 1 \\ -3 \end{bmatrix} = \begin{bmatrix} k_1 - k_2 \\ 2k_2 + k_3 \\ -k_1 + 2k_2 + 2k_3 \end{bmatrix}.$$

$$k_1 - k_2 = 2, \ 2k_2 + k_3 = 1, -k_1 + 2k_2 + 2k_3 = -3.$$

Using Gauss elimination method,

$$k_1 = 3, k_2 = 1, k_3 = -1$$

$$T(v_2) = 3w_1 + w_2 - w_3$$

$$\left[T(v_2)\right]_{s_2} = \begin{bmatrix} 3 \\ 1 \\ -1 \end{bmatrix}.$$

The matrix of the transformation T with respect to the bases S_1 and S_2 is

$$[T]_{s_2,s_1} = \left[\left[T(v_1)\right]_{s_2} | \left[T(v_2)\right]_{s_2}\right] = \begin{bmatrix} 1 & 3 \\ 0 & 1 \\ -2 & -1 \end{bmatrix}.$$

Example 4.2

Let $T : P_2 \to P_2$ be the linear operator defined by $T(p(x)) = p(2x+1)$. Find the matrix of the operator T with respect to the bases $S = \{p_1, p_2, p_3\}$, where $p_1 = 1, p_2 = x, p_3 = x^2$. Also, compute $T(2 - 3x + 4x^2)$.

Solution

Here, $T(p_1) = T(1) = 1$

$$T(p_2) = T(x) = 2x + 1,$$

$$T(p_3) = T(x^2) = (2x+1)^2 = 4x^2 + 4x + 1.$$

Let us express p_1 as a linear combination of p_1, p_2, p_3

$$T(p_1) = k_1 p_1 + k_2 p_2 + k_3 p_3$$

$$\therefore 1 = k_1 1 + k_2 x + k_3 x^2.$$

Comparing both sides, coefficient of $1, x, x^2$

$$k_1 = 1, k_2 = 0, k_3 = 0$$

$$T(p_1) = 1$$

$$[T(p_1)]_s = \begin{bmatrix} 1 \\ 0 \\ 0 \end{bmatrix}.$$

Express p_2 as a linear combination of p_1, p_2, p_3

$$T(p_2) = k_1 p_1 + k_2 p_2 + k_3 p_3$$

$$\therefore 2x + 1 = k_1 1 + k_2 x + k_3 x^2.$$

Comparing both sides, coefficient of $1, x, x^2$

$$k_1 = 1, k_2 = 2, k_3 = 0$$

$$T(p_2) = p_1 + 2p_2$$

$$[T(p_2)]_s = \begin{bmatrix} 1 \\ 2 \\ 0 \end{bmatrix}.$$

Express p_3 as a linear combination of p_1, p_2, p_3

$$T(p_3) = k_1 p_1 + k_2 p_2 + k_3 p_3$$

$$\therefore 4x^2 + 4x + 1 = k_1 1 + k_2 x + k_3 x^2.$$

Comparing both sides, coefficient of $1, x, x^2$

$$k_1 = 1, k_2 = 4, k_3 = 4$$

$$T(p_2) = p_1 + 4p_2 + 4p_3$$

$$\left[T(p_3)\right]_s = \begin{bmatrix} 1 \\ 4 \\ 4 \end{bmatrix}.$$

The matrix of the operator T with respect to the basis S is

$$[T]_s = \left[\left[T(p_1)\right]_s \mid \left[T(p_2)\right]_s \mid \left[T(p_3)\right]_s\right] = \begin{bmatrix} 1 & 1 & 1 \\ 0 & 2 & 4 \\ 0 & 0 & 4 \end{bmatrix}.$$

Example 4.3

Let $T : M_{22} \to M_{22}$ be the linear operator defined by $T(A) = A^T$. Find the matrix of the operator T with respect to the bases $S_1 = \{A_1, A_2, A_3, A_4\}$ and $S_2 = \{B_1, B_2, B_3, B_4\}$, where $A_1 = \begin{bmatrix} 1 & 0 \\ 0 & 0 \end{bmatrix}$, $A_2 = \begin{bmatrix} 0 & 1 \\ 0 & 0 \end{bmatrix}$, $A_3 = \begin{bmatrix} 0 & 0 \\ 1 & 0 \end{bmatrix}$, $A_4 = \begin{bmatrix} 0 & 0 \\ 0 & 1 \end{bmatrix}$, $B_1 = \begin{bmatrix} 1 & 1 \\ 0 & 0 \end{bmatrix}$, $B_2 = \begin{bmatrix} 0 & 1 \\ 0 & 0 \end{bmatrix}$, $B_3 = \begin{bmatrix} 0 & 0 \\ 1 & 1 \end{bmatrix}$, $B_4 = \begin{bmatrix} 1 & 0 \\ 0 & 1 \end{bmatrix}$.

Solution

Here, $T(A_1) = T\left(\begin{bmatrix} 1 & 0 \\ 0 & 0 \end{bmatrix}\right) = \begin{bmatrix} 1 & 0 \\ 0 & 0 \end{bmatrix}^T = \begin{bmatrix} 1 & 0 \\ 0 & 0 \end{bmatrix}$

$$T(A_2) = T\left(\begin{bmatrix} 0 & 1 \\ 0 & 0 \end{bmatrix}\right) = \begin{bmatrix} 0 & 1 \\ 0 & 0 \end{bmatrix}^T = \begin{bmatrix} 0 & 0 \\ 1 & 0 \end{bmatrix}$$

$$T(A_3) = T\left(\begin{bmatrix} 0 & 0 \\ 1 & 0 \end{bmatrix}\right) = \begin{bmatrix} 0 & 0 \\ 1 & 0 \end{bmatrix}^T = \begin{bmatrix} 0 & 1 \\ 0 & 0 \end{bmatrix}$$

$$T(A_4) = T\left(\begin{bmatrix} 0 & 0 \\ 0 & 1 \end{bmatrix}\right) = \begin{bmatrix} 0 & 0 \\ 0 & 1 \end{bmatrix}^T = \begin{bmatrix} 0 & 0 \\ 0 & 1 \end{bmatrix}.$$

Let us express A_1 as a linear combination of B_1, B_2, B_3, B_4

$$T(A_1) = k_1 B_1 + k_2 B_2 + k_3 B_3 + k_4 B_4$$

$$\therefore \begin{bmatrix} 1 & 0 \\ 0 & 0 \end{bmatrix} = k_1 \begin{bmatrix} 1 & 1 \\ 0 & 0 \end{bmatrix} + k_2 \begin{bmatrix} 0 & 1 \\ 0 & 0 \end{bmatrix} + k_3 \begin{bmatrix} 0 & 0 \\ 1 & 1 \end{bmatrix} + k_4 \begin{bmatrix} 1 & 0 \\ 0 & 1 \end{bmatrix}.$$

$$\therefore \begin{bmatrix} 1 & 0 \\ 0 & 0 \end{bmatrix} = \begin{bmatrix} k_1 + k_4 & k_1 + k_2 \\ k_3 & k_3 + k_4 \end{bmatrix}.$$

Comparing both sides,

$$k_1 + k_4 = 1, k_1 + k_2 = 0, k_3 = 0, k_3 + k_4 = 0.$$

Using Gauss elimination, we get

$$k_1 = 1, k_2 = -1, k_3 = 0, k_4 = 0$$

$$\therefore T(A_1) = B_1 - B_2$$

$$[T(A_1)]_{s_1} = \begin{bmatrix} 1 \\ -1 \\ 0 \\ 0 \end{bmatrix}.$$

Let us express A_2 as a linear combination of B_1, B_2, B_3, B_4

$$T(A_2) = k_1 B_1 + k_2 B_2 + k_3 B_3 + k_4 B_4.$$

$$\therefore \begin{bmatrix} 0 & 0 \\ 1 & 0 \end{bmatrix} = k_1 \begin{bmatrix} 1 & 1 \\ 0 & 0 \end{bmatrix} + k_2 \begin{bmatrix} 0 & 1 \\ 0 & 0 \end{bmatrix} + k_3 \begin{bmatrix} 0 & 0 \\ 1 & 1 \end{bmatrix} + k_4 \begin{bmatrix} 1 & 0 \\ 0 & 1 \end{bmatrix}$$

$$\therefore \begin{bmatrix} 0 & 0 \\ 1 & 0 \end{bmatrix} = \begin{bmatrix} k_1 + k_4 & k_1 + k_2 \\ k_3 & k_3 + k_4 \end{bmatrix}.$$

Comparing both sides,

$$k_1 + k_4 = 0, k_1 + k_2 = 0, k_3 = 1, k_3 + k_4 = 0.$$

Using Gauss elimination, we get

$$k_1 = 1, k_2 = -1, k_3 = 1, k_4 = -1$$

$$\therefore T(A_2) = B_1 - B_2 + B_3 - B_4$$

$$[T(A_2)]_{s_1} = \begin{bmatrix} 1 \\ -1 \\ 1 \\ -1 \end{bmatrix}.$$

Let us express A_3 as a linear combination of B_1, B_2, B_3, B_4

$$T(A_3) = k_1 B_1 + k_2 B_2 + k_3 B_3 + k_4 B_4$$

$$\therefore \begin{bmatrix} 0 & 1 \\ 0 & 0 \end{bmatrix} = k_1 \begin{bmatrix} 1 & 1 \\ 0 & 0 \end{bmatrix} + k_2 \begin{bmatrix} 0 & 1 \\ 0 & 0 \end{bmatrix} + k_3 \begin{bmatrix} 0 & 0 \\ 1 & 1 \end{bmatrix} + k_4 \begin{bmatrix} 1 & 0 \\ 0 & 1 \end{bmatrix}$$

$$\therefore \begin{bmatrix} 0 & 1 \\ 0 & 0 \end{bmatrix} = \begin{bmatrix} k_1 + k_4 & k_1 + k_2 \\ k_3 & k_3 + k_4 \end{bmatrix}.$$

Comparing both sides,

$$k_1 + k_4 = 0, k_1 + k_2 = 1, k_3 = 0, k_3 + k_4 = 0.$$

Using Gauss elimination, we get

$$k_1 = 0, k_2 = 1, k_3 = 0, k_4 = 0$$

$$\therefore T(A_3) = B_2$$

$$\left[T(A_3)\right]_{s_1} = \begin{bmatrix} 0 \\ 1 \\ 0 \\ 0 \end{bmatrix}.$$

Let us express A_4 as a linear combination of B_1, B_2, B_3, B_4

$$T(A_4) = k_1 B_1 + k_2 B_2 + k_3 B_3 + k_4 B_4$$

$$\therefore \begin{bmatrix} 0 & 0 \\ 0 & 1 \end{bmatrix} = k_1 \begin{bmatrix} 1 & 1 \\ 0 & 0 \end{bmatrix} + k_2 \begin{bmatrix} 0 & 1 \\ 0 & 0 \end{bmatrix} + k_3 \begin{bmatrix} 0 & 0 \\ 1 & 1 \end{bmatrix} + k_4 \begin{bmatrix} 1 & 0 \\ 0 & 1 \end{bmatrix}$$

$$\therefore \begin{bmatrix} 0 & 0 \\ 0 & 1 \end{bmatrix} = \begin{bmatrix} k_1 + k_4 & k_1 + k_2 \\ k_3 & k_3 + k_4 \end{bmatrix}.$$

Comparing both sides,

$$k_1 + k_4 = 0, k_1 + k_2 = 0, k_3 = 0, k_3 + k_4 = 1.$$

Using the Gauss elimination, we get

$$k_1 = -1, k_2 = 1, k_3 = 0, k_4 = 1$$

$$\therefore T(A_4) = -B_1 + B_2 + B_4$$

$$[T(A_2)]_{S_1} = \begin{bmatrix} -1 \\ 1 \\ 0 \\ 1 \end{bmatrix}.$$

The matrix of the transformation T with respect to the bases S_1 and S_2 is

$$[T]_{S_2,S_1} = \left[[T(A_1)]_{S_2} \ | \ [T(A_2)]_{S_2} \ | \ [T(A_3)]_{S_2} \ | \ [T(A_4)]_{S_2} \right] = \begin{bmatrix} 1 & 1 & 0 & -1 \\ -1 & -1 & 1 & 1 \\ 0 & 1 & 0 & 0 \\ 0 & -1 & 0 & 1 \end{bmatrix}.$$

Example 4.4

Let $T: R^2 \to R^2$ be the linear operator defined by $T\left(\begin{bmatrix} x_1 \\ x_2 \end{bmatrix} \right) = \begin{bmatrix} x_1 + x_2 \\ -2x_1 + 4x_2 \end{bmatrix}.$

Find the matrix of the operator T with respect to the basis $S = \{u_1, u_2\}$, where $u_1 = \begin{bmatrix} 1 \\ 1 \end{bmatrix}$ and $u_2 = \begin{bmatrix} 1 \\ 2 \end{bmatrix}$. Also, verify $[T]_S [X]_S = [T(X)]_S.$

Solution

Here, $T(u_1) = T\left(\begin{bmatrix} 1 \\ 1 \end{bmatrix} \right) = \begin{bmatrix} 1+1 \\ -2+4 \end{bmatrix} = \begin{bmatrix} 2 \\ 2 \end{bmatrix}$

$$T(u_2) = T\left(\begin{bmatrix} 1 \\ 2 \end{bmatrix} \right) = \begin{bmatrix} 1+2 \\ -2+8 \end{bmatrix} = \begin{bmatrix} 3 \\ 6 \end{bmatrix}.$$

Let us express u_1 as a linear combination of u_1, u_2

$$T(u_1) = k_1 u_1 + k_2 u_2$$

$$\therefore \begin{bmatrix} 2 \\ 2 \end{bmatrix} = k_1 \begin{bmatrix} 1 \\ 1 \end{bmatrix} + k_2 \begin{bmatrix} 1 \\ 2 \end{bmatrix}$$

$$\therefore \begin{bmatrix} 2 \\ 2 \end{bmatrix} = \begin{bmatrix} k_1 + k_2 \\ k_1 + 2k_2 \end{bmatrix}.$$

Comparing both sides,

$$k_1 + k_2 = 2, k_1 + 2k_2 = 2.$$

Using Gauss elimination method,

$$k_1 = 2, k_2 = 0$$

$$T(u_1) = 2u_1$$

$$\left[T(u_1)\right]_s = \begin{bmatrix} 2 \\ 0 \end{bmatrix}.$$

Express u_2 as a linear combination of u_1, u_2

$$T(u_2) = k_1 u_1 + k_2 u_2$$

$$\therefore \begin{bmatrix} 3 \\ 6 \end{bmatrix} = k_1 \begin{bmatrix} 1 \\ 1 \end{bmatrix} + k_2 \begin{bmatrix} 1 \\ 2 \end{bmatrix}$$

$$\therefore \begin{bmatrix} 3 \\ 6 \end{bmatrix} = \begin{bmatrix} k_1 + k_2 \\ k_1 + 2k_2 \end{bmatrix}$$

$$k_1 + k_2 = 3, \ k_1 + 2k_2 = 6.$$

Using Gauss elimination method,

$$k_1 = 0, k_2 = 3$$

$$T(u_2) = 3u_2$$

$$\left[T(u_2)\right]_s = \begin{bmatrix} 0 \\ 3 \end{bmatrix}.$$

The matrix of the transformation T with respect to the basis S is

$$[T]_s = \left[\left[T(u_1)\right]_s \mid \left[T(u_2)\right]_s\right] = \begin{bmatrix} 2 & 0 \\ 0 & 3 \end{bmatrix}.$$

If $X = \begin{bmatrix} x_1 \\ x_2 \end{bmatrix}$ is any vector in R^2, then the formula for $T(X) = \begin{bmatrix} x_1 + x_2 \\ -2x_1 + 4x_2 \end{bmatrix}.$

To find $\left[T(X)\right]_s$ and $[X]_s$, we must express $\left[T(X)\right]_s$ and $[X]_s$ as a linear combination of u_1 and u_2.

Therefore, $X = k_1 u_1 + k_2 u_2$

$$\therefore \begin{bmatrix} x_1 \\ x_2 \end{bmatrix} = k_1 \begin{bmatrix} 1 \\ 1 \end{bmatrix} + k_2 \begin{bmatrix} 1 \\ 2 \end{bmatrix}$$

$$\therefore k_1 + k_2 = x_1, \ k_1 + 2k_2 = x_2$$

$$\therefore k_1 = 2x_1 - x_2, \ k_2 = -x_1 + x_2$$

$$\therefore [X]_S = \begin{bmatrix} 2x_1 - x_2 \\ -x_1 + x_2 \end{bmatrix}$$

$$\therefore [T]_S [X]_S = \begin{bmatrix} 2 & 0 \\ 0 & 3 \end{bmatrix} \begin{bmatrix} 2x_1 - x_2 \\ -x_1 + x_2 \end{bmatrix} = \begin{bmatrix} 4x_1 - 2x_2 \\ -3x_1 + 3x_2 \end{bmatrix} \tag{4.2}$$

For $[T(X)]_{S'}$

$$T(X) = c_1 u_1 + c_2 u_2$$

$$\therefore \begin{bmatrix} x_1 + x_2 \\ -2x_1 + 4x_2 \end{bmatrix} = c_1 \begin{bmatrix} 1 \\ 1 \end{bmatrix} + c_2 \begin{bmatrix} 1 \\ 2 \end{bmatrix}$$

$$\therefore \begin{bmatrix} x_1 + x_2 \\ -2x_1 + 4x_2 \end{bmatrix} = \begin{bmatrix} c_1 + c_2 \\ c_1 + 2c_2 \end{bmatrix}$$

$$\therefore c_1 + c_2 = x_1 + x_2, \ c_1 + 2c_2 = -2x_1 + 4x_2.$$

To find c_1, c_2, use Gauss elimination.
So, $c_1 = 4x_1 - 2x_2$ and $c_2 = -3x_1 + 3x_2$.

Thus,
$$[T(X)]_S = \begin{bmatrix} 4x_1 - 2x_2 \\ -3x_1 + 3x_2 \end{bmatrix}. \tag{4.3}$$

From equations (4.2) and (4.3),

$$\therefore [T]_S [X]_S = [T(X)]_S.$$

Hence, it is verified.

Example 4.5

Let $T : P_1 \to P_1$ be defined by $T(a_0 + a_1 x) = a_0 + a_1(x+1)$, $S_1 = \{p_1, p_2\}$, and $S_2 = \{q_1, q_2\}$ where $p_1 = 6 + 3x$, $p_2 = 10 + 2x$, $q_1 = 10 + 2x$, and $q_2 = 3 + 2x$. Find the matrix of T with respect to the basis S_1 and matrix of T with respect to the basis S_2.

Solution

$T(p_1) = T(6+3x) = 6+3(x+1) = 9+3x.$

Expressing $T(p_1)$ as linear combinations of p_1 and p_2

$$T(p_1) = k_1 p_1 + k_2 p_2$$

$$9+3x = k_1(6+3x) + k_2(10+2x)$$

$$= (6k_1 + 10k_2) + (3k_1 + 2k_2)x.$$

Equating corresponding coefficients,

$$6k_1 + 10k_2 = 9, \quad 3k_1 + 2k_2 = 3.$$

Solving these equations,

$$k_1 = \frac{2}{3}, k_2 = \frac{1}{2}$$

$$\therefore T(p_1) = \frac{2}{3} p_1 + \frac{1}{2} p_2$$

$$\therefore \left[T(p_1) \right]_{S_1} = \begin{bmatrix} \dfrac{2}{3} \\ \dfrac{1}{2} \end{bmatrix}.$$

Similarly, $T(p_2) = T(10+2x) = 10+2(x+1) = 12+2x.$

Expressing $T(p_2)$ as linear combinations of p_1 and p_2

$$T(p_2) = k_1 p_1 + k_2 p_2$$

$$12+2x = k_1(6+3x) + k_2(10+2x)$$

$$= (6k_1 + 10k_2) + (3k_1 + 2k_2)x.$$

Equating corresponding coefficients,

$$6k_1 + 10k_2 = 12, \quad 3k_1 + 2k_2 = 2.$$

Solving these equations,

$$k_1 = -\frac{2}{9}, k_2 = \frac{4}{3}$$

$$\therefore T(p_2) = -\frac{2}{9} p_1 + \frac{4}{3} p_2$$

$$\therefore \left[T(p_2) \right]_{S_1} = \begin{bmatrix} -\dfrac{2}{9} \\ \dfrac{4}{3} \end{bmatrix}$$

$$\therefore \left[T \right]_{S_1} = \left[\left[T(p_1) \right]_{S_1} \middle| \left[T(p_2) \right]_{S_1} \right]$$

$$= \begin{bmatrix} \dfrac{2}{3} & -\dfrac{2}{9} \\ \dfrac{1}{2} & \dfrac{4}{3} \end{bmatrix}.$$

Expressing q_1 as linear combinations of p_1 and p_2

$$q_1 = k_1 p_1 + k_2 p_2$$

$$2 = k_1(6+3x) + k_2(10+2x)$$

$$= (6k_1 + 10k_2) + (3k_1 + 2k_2)x.$$

Equating corresponding coefficients,

$$6k_1 + 10k_2 = 2, \quad 3k_1 + 2k_2 = 0.$$

Solving these equations,

$$k_1 = -\frac{2}{9}, k_2 = \frac{1}{3}$$

$$\therefore q_1 = -\frac{2}{9}p_1 + \frac{1}{3}p_2$$

$$\therefore \left[q_1 \right]_{S_1} = \begin{bmatrix} -\dfrac{2}{9} \\ \dfrac{1}{3} \end{bmatrix}.$$

Expressing q_2 as linear combinations of p_1 and p_2

$$q_2 = k_1 p_1 + k_2 p_2$$

$$3 + 2x = k_1(6+3x) + k_2(10+2x)$$

$$= (6k_1 + 10k_2) + (3k_1 + 2k_2)x.$$

Equating corresponding coefficients,

$$6k_1 + 10k_2 = 3, \quad 3k_1 + 2k_2 = 2.$$

Solving these equations,

$$k_1 = \frac{7}{9}, k_2 = -\frac{1}{6}$$

$$\therefore q_2 = \frac{7}{9} p_1 - \frac{1}{6} p_2$$

$$\therefore [q_2]_{S_1} = \begin{bmatrix} \frac{7}{9} \\ -\frac{1}{6} \end{bmatrix}.$$

Hence, the transition matrix from S_2 to S_1 is

$$P = \left[[q_1]_{S_1} \middle| [q_2]_{S_1} \right]$$

$$P = \begin{bmatrix} -\frac{2}{9} & \frac{7}{9} \\ \frac{1}{3} & -\frac{1}{6} \end{bmatrix}.$$

Thus, $P^{-1} = \begin{bmatrix} \frac{3}{4} & \frac{7}{2} \\ \frac{3}{2} & 1 \end{bmatrix}.$

$$[T]_{S_2} = P^{-1}[T]_{S_1} P = \begin{bmatrix} \frac{3}{4} & \frac{7}{2} \\ \frac{3}{2} & 1 \end{bmatrix} \begin{bmatrix} \frac{2}{3} & -\frac{2}{9} \\ \frac{1}{2} & \frac{4}{3} \end{bmatrix} \begin{bmatrix} -\frac{2}{9} & \frac{7}{9} \\ \frac{1}{3} & -\frac{1}{6} \end{bmatrix}.$$

$$= \begin{bmatrix} 1 & 1 \\ 0 & 1 \end{bmatrix}$$

In the next section, we will discuss the similarity of the linear matrix transformation and its examples.

4.2 SIMILARITY

If P and Q are two square matrices, then Q is said to be similar to P, if there exists a non-singular matrix A such that $Q = A^{-1}PA$.

Properties
If P and Q are similar matrices, then

1. P and Q have the same determinant.
2. P and Q have the same rank.

3. P and Q have the same nullity.
4. P and Q have the same trace.
5. P and Q have the same characteristic polynomial.
6. P and Q have the same eigenvalues.
7. If λ is an eigenvalue of two similar matrices, the eigenspace of both the similar matrices corresponding to λ has the same dimension.

Example 4.6

Show that the matrices $\begin{bmatrix} 1 & 1 \\ -1 & 4 \end{bmatrix}$ and $\begin{bmatrix} 2 & 1 \\ 1 & 3 \end{bmatrix}$ are similar but that

$\begin{bmatrix} 3 & 1 \\ -6 & -2 \end{bmatrix}$ and $\begin{bmatrix} -1 & 2 \\ 1 & 0 \end{bmatrix}$ are not.

Solution

Let $A_1 = \begin{bmatrix} 1 & 1 \\ -1 & 4 \end{bmatrix}$, $A_2 = \begin{bmatrix} 2 & 1 \\ 1 & 3 \end{bmatrix}$, $A_3 = \begin{bmatrix} 3 & 1 \\ -6 & -2 \end{bmatrix}$, and $A_4 = \begin{bmatrix} -1 & 2 \\ 1 & 0 \end{bmatrix}$.

$$\det(A_1) = \begin{vmatrix} 1 & 1 \\ -1 & 4 \end{vmatrix} = 5, \ \det(A_2) = \begin{vmatrix} 2 & 1 \\ 1 & 3 \end{vmatrix} = 5$$

$$\det(A_3) = \begin{vmatrix} 3 & 1 \\ -6 & -2 \end{vmatrix} = 0, \ \det(A_4) = \begin{vmatrix} -1 & 2 \\ 1 & 0 \end{vmatrix} = -2.$$

Therefore, if $\det(A_1) = \det(A_2)$, matrices A_1 and A_2 are similar.
If $\det(A_3) \neq \det(A_4)$, matrices A_1 and A_2 are not similar.

4.2.1 EFFECT OF CHANGING BASES ON MATRICES OF LINEAR OPERATORS

In this section, we will discuss the effect on the transformation if we change different bases.

If S_1 and S_2 are two bases for a finite-dimensional vector space V and if $T : V \rightarrow V$ is the identity operator, what is the relationship between $[T]_{S_1}$ and $[T]_{S_2}$ if it exists?

Theorem 4.1

Let $T : V \rightarrow V$ be a linear operator on a finite-dimensional vector space V, and let S_1 and S_2 be bases for V. Then, $[T]_{S_2} = P^{-1}[T]_{S_1} P$, where P is the transition matrix from S_2 to S_1.

Example 4.7

Let $T: R^2 \rightarrow R^2$ be defined by $T\left(\begin{bmatrix} x_1 \\ x_2 \end{bmatrix}\right) = \begin{bmatrix} x_1 + x_2 \\ -2x_1 + 4x_2 \end{bmatrix}$. Find the matrix for T with respect to the basis $S_1 = \{e_1, e_2\}$ for R^2, and also find a matrix for T with respect to the basis $S_2 = \{u_1', u_2'\}$ for R^2, where $u_1' = \begin{bmatrix} 1 \\ 1 \end{bmatrix}$ and $u_2' = \begin{bmatrix} 1 \\ 2 \end{bmatrix}$.

Solution

$$T(e_1) = T\left(\begin{bmatrix} 1 \\ 0 \end{bmatrix}\right) = \begin{bmatrix} 1+0 \\ -2+0 \end{bmatrix} = \begin{bmatrix} 1 \\ -2 \end{bmatrix}$$

$$T(e_2) = T\left(\begin{bmatrix} 0 \\ 1 \end{bmatrix}\right) = \begin{bmatrix} 0+1 \\ 0+4 \end{bmatrix} = \begin{bmatrix} 1 \\ 4 \end{bmatrix}.$$

Let us express e_1 as a linear combination of e_1, e_2

$$T(e_1) = k_1 e_1 + k_2 e_2$$

$$\therefore \begin{bmatrix} 1 \\ -2 \end{bmatrix} = k_1 \begin{bmatrix} 1 \\ 0 \end{bmatrix} + k_2 \begin{bmatrix} 0 \\ 1 \end{bmatrix}$$

$$\therefore \begin{bmatrix} 1 \\ -2 \end{bmatrix} = \begin{bmatrix} k_1 \\ k_2 \end{bmatrix}.$$

Comparing both sides,

$$k_1 = 1, \ k_2 = -2$$

$$T(e_1) = u_1 - 2u_2$$

$$[T(e_1)]_{s_1} = \begin{bmatrix} 1 \\ -2 \end{bmatrix}.$$

Express e_2 as a linear combination of e_1, e_2

$$T(e_2) = k_1 e_1 + k_2 e_2$$

$$\therefore \begin{bmatrix} 1 \\ 4 \end{bmatrix} = k_1 \begin{bmatrix} 1 \\ 0 \end{bmatrix} + k_2 \begin{bmatrix} 0 \\ 1 \end{bmatrix}$$

$$\therefore \begin{bmatrix} 1 \\ 4 \end{bmatrix} = \begin{bmatrix} k_1 \\ k_2 \end{bmatrix}$$

$$k_1 = 1, k_2 = 4$$

$$T(e_2) = u_1 + 4u_2$$

$$[T(e_2)]_{s_1} = \begin{bmatrix} 1 \\ 4 \end{bmatrix}.$$

The matrix of the transformation T with respect to the basis S_1 is

$$[T]_{S_1} = \left[[T(e_1)]_{s_1} \mid [T(e_2)]_{s_1} \right] = \begin{bmatrix} 1 & 1 \\ -2 & 4 \end{bmatrix}.$$

To find $[T]_{S_2}$

$$u_1' = \begin{bmatrix} 1 \\ 1 \end{bmatrix}, \ u_2' = \begin{bmatrix} 1 \\ 2 \end{bmatrix}$$

$$T(u_1') = T\left(\begin{bmatrix} 1 \\ 1 \end{bmatrix} \right) = \begin{bmatrix} 1+1 \\ -2+4 \end{bmatrix} = \begin{bmatrix} 2 \\ 2 \end{bmatrix}$$

$$T(u_2') = T\left(\begin{bmatrix} 1 \\ 2 \end{bmatrix} \right) = \begin{bmatrix} 1+2 \\ -2+8 \end{bmatrix} = \begin{bmatrix} 3 \\ 6 \end{bmatrix}.$$

Let us express u_1' as a linear combination of u_1', u_2'

$$T(u_1') = k_1 u_1' + k_2 u_2'$$

$$\therefore \begin{bmatrix} 2 \\ 2 \end{bmatrix} = k_1 \begin{bmatrix} 1 \\ 1 \end{bmatrix} + k_2 \begin{bmatrix} 1 \\ 2 \end{bmatrix}$$

$$\therefore \begin{bmatrix} 2 \\ 2 \end{bmatrix} = \begin{bmatrix} k_1 + k_2 \\ k_1 + 2k_2 \end{bmatrix}.$$

Comparing both sides,

$$k_1 + k_2 = 2, \ k_1 + 2k_2 = 2$$

$$k_1 = 2, \ k_2 = 0$$

$$T(u_2') = 2u_1$$

$$[T(u_2')]_{s_2} = \begin{bmatrix} 2 \\ 0 \end{bmatrix}.$$

Let us express u_2' as a linear combination of u_1', u_2'

$$T\left(u_2'\right) = k_1 u_1' + k_2 u_2'$$

$$\therefore \begin{bmatrix} 3 \\ 6 \end{bmatrix} = k_1 \begin{bmatrix} 1 \\ 1 \end{bmatrix} + k_2 \begin{bmatrix} 1 \\ 2 \end{bmatrix}$$

$$\therefore \begin{bmatrix} 3 \\ 6 \end{bmatrix} = \begin{bmatrix} k_1 + k_2 \\ k_1 + 2k_2 \end{bmatrix}.$$

Comparing both sides,

$$k_1 + k_2 = 3, k_1 + 2k_2 = 6$$

$$k_1 = 0, \quad k_2 = 3$$

$$T\left(u_2'\right) = 3u_2$$

$$\left[T\left(u_2'\right)\right]_{S_2} = \begin{bmatrix} 0 \\ 3 \end{bmatrix}.$$

The matrix of the transformation T with respect to the basis S_2 is

$$[T]_{S_2} = \left[\left[T\left(u_1'\right)\right]_{S_2} \mid \left[T\left(u_2'\right)\right]_{S_2}\right] = \begin{bmatrix} 2 & 0 \\ 0 & 3 \end{bmatrix}.$$

EXERCISE SET 4

Q.1 Find the matrix for the linear operator $T: R^2 \to R^2$ which is defined by $T\left(\begin{bmatrix} x_1 \\ x_2 \end{bmatrix}\right) = \begin{bmatrix} x_1 - 2x_2 \\ -x_2 \end{bmatrix}$ relative to the bases $S_1 = \{u_1, u_2\}$ and $S_2 = \{v_1, v_2\}$, where $u_1 = \begin{bmatrix} 1 \\ 0 \end{bmatrix}, u_1 = \begin{bmatrix} 0 \\ 1 \end{bmatrix}$ and $v_1 = \begin{bmatrix} 4 \\ 1 \end{bmatrix}, v_1 = \begin{bmatrix} 7 \\ 2 \end{bmatrix}.$

Q.2 Find the matrix for the linear transformation $T : R^2 \rightarrow R^3$ which is defined

by $T\left(\begin{bmatrix} x_1 \\ x_2 \end{bmatrix}\right) = \begin{bmatrix} 3x_1 + 2x_2 \\ -x_2 \\ x_1 \end{bmatrix}$ relative to the bases $S_1 = \{u_1, u_2\}$ and

$S_2 = \{v_1, v_2\}$, where $u_1 = \begin{bmatrix} -1 \\ 2 \end{bmatrix}$, $u_1 = \begin{bmatrix} 3 \\ 4 \end{bmatrix}$ and $v_1 = \begin{bmatrix} 4 \\ 1 \\ 3 \end{bmatrix}$, $v_2 = \begin{bmatrix} 7 \\ 0 \\ 1 \end{bmatrix}$,

$v_3 = \begin{bmatrix} 1 \\ 1 \\ 0 \end{bmatrix}$.

Q.3 Find the matrix for the linear transformation $T : R^2 \rightarrow M_{22}$ which is defined by
$T\left(\begin{bmatrix} x_1 \\ x_2 \end{bmatrix}\right) = \begin{bmatrix} x_1 - 4x_2 & 2x_1 + x_2 \\ x_2 & 5x_1 - 7x_2 \end{bmatrix}$ relative to the bases $S_1 = \{u_1, u_2\}$

and $S_2 = \{v_1, v_2\}$, where $u_1 = \begin{bmatrix} 1 \\ 0 \end{bmatrix}$, $u_2 = \begin{bmatrix} 0 \\ 1 \end{bmatrix}$ and $v_1 = \begin{bmatrix} 4 & 0 \\ 0 & 1 \end{bmatrix}$,

$v_2 = \begin{bmatrix} 2 & -1 \\ 3 & 5 \end{bmatrix}$, $v_2 = \begin{bmatrix} 3 & -1 \\ 1 & 2 \end{bmatrix}$, $v_2 = \begin{bmatrix} 1 & -1 \\ 0 & 2 \end{bmatrix}$.

Q.4 Find the matrix for the linear transformation $T : P_2 \rightarrow P_3$ which is defined by
$T(p(x)) = xp(x)$ relative to the bases $S_1 = \{u_1, u_2, u_3\}$ and $S_2 = \{v_1, v_2, v_3, v_4\}$,
where $u_1 = 1$, $u_2 = x$, $u_3 = x^2$ and $v_1 = 1$, $v_2 = x$, $v_3 = x^2$, $v_4 = x^3$.

Q.5 Find the matrix for the linear transformation $T : P_2 \rightarrow P_1$ which is defined
by $T(a_0 + a_1x + a_2x^2) = (a_0 + a_1) - (2a_1 + 3a_2)x$ relative to the bases
$S_1 = \{u_1, u_2, u_3\}$ and $S_2 = \{v_1, v_2\}$, where, $u_1 = 1$, $u_2 = x$, $u_3 = x^2$ and $v_1 = 1$,
$v_2 = x$. Also, obtain $T(3 + x - x^2)$.

ANSWERS TO EXERCISE SET 4

Q.1 $[T]_S = \begin{bmatrix} 2 & 3 \\ -1 & -2 \end{bmatrix}$

Q.2 $[T]_{S_1,S_2} = \begin{bmatrix} -\dfrac{5}{9} & \dfrac{4}{9} \\[2mm] \dfrac{2}{3} & \dfrac{5}{3} \\[2mm] -\dfrac{13}{9} & \dfrac{32}{9} \end{bmatrix}$

Q.3 $[T]_{S_1,S_2} = \begin{bmatrix} \dfrac{54}{17} & -\dfrac{40}{17} \\[2mm] \dfrac{33}{17} & -\dfrac{15}{17} \\[2mm] -\dfrac{99}{17} & \dfrac{62}{17} \\[2mm] \dfrac{32}{17} & -\dfrac{64}{17} \end{bmatrix}$

Q.4 $[T]_{S_1,S_2} = \begin{bmatrix} 0 & 0 & 0 \\ 1 & 0 & 0 \\ 0 & 1 & 0 \\ 0 & 0 & 1 \end{bmatrix}$

Q.5 $[T]_{S_1,S_2} = \begin{bmatrix} 1 & 1 & 0 \\ 0 & -2 & -3 \end{bmatrix}, T\left(3+x-x^2\right) = 4+x$

Multiple-Choice Questions on Linear Transformation

1   Let V and W be the vector space over R and $T : V \to W$ be a linear map, then

(a) T is one-one if T maps linearly independent subset of V to linear independent subset of W

(b) T is one-one if $\operatorname{Ker}(T)$ is a proper subset of V

(c) T is one-one if $\dim(V) \leq \dim(W)$

(d) T is one-one and maps a spanning set of V to a spanning set of W

2   Let $\langle,\rangle$ be an inner product on R^4, v be a nonzero vector in R^4, and $T : R^4 \to R$ be defined as $T(x) = \langle x, v \rangle$ for all $x \in R^4$. Then, it is not true that

(a) T is a linear map

(b) T is a linear map and one-one

(c) T is a linear map and $\dim\left(\ker(T)\right) = 3$

(d) T is onto

3   Let V be a real vector space and $S, T : V \to V$ be the linear maps such that $ST = TS$. Then

(a) $\ker(S)$ is invariant under T

(b) $\ker(S) = \ker(T)$

(c) $\operatorname{Im}(S) = \operatorname{Im}(T)$

(d) S and T are similar

4   Let $p_1(R)$ be the vector space of all real polynomials of degree ≤ 1 and $T : p_1(R) \to R^2$ be the linear map defined by $T\left(p(x)\right) = \left(p(0), p(1)\right)$. Then,

(a) T is one-one but not onto

(b) T is onto but not one-one

(c) T is neither one-one nor onto

(d) T is one-one and onto

5   The transformation $T(x, y) = (x, y, x - y)$ is?

(a) Onto

(b) One-one

(c) Both

(d) None of these

6   If $\{e_1, e_2, e_3\}$ is a standard basis of R^3 and $T : R^3 \to R$ be a linear transformation such that $T(e_3) = 2e_1 + 3e_2 + 5e_3$, $T(e_2 + e_3) = e_1$, $T(e_1 + e_2 + e_3) = e_2 - e_3$, then $T(e_1 + 2e_2 + 3e_3) = ?$

(a) $(1, 4, 4)$

(b) $(2, 3, 4)$

(c) $(3, 4, 4)$

(d) $(1, 2, 3)$

7   Let $T : R^4 \to R^3$ be a linear transformation with $R(T) = 3$, then what is the nullity of T?

(a) 2

(b) 1

(c) 3

(d) 4

8   Let $T : R^3 \to R^3$ be one-to-one linear transformation, then the dimension of $\ker(T)$ is

(a) 1

(b) 2

(c) 0

(d) 3

9   Let $T : R^2 \to R^2$ be a linear transformation defined by $T(x, y) = (y, x)$, then it is

(a) Onto

(b) One-one

(c) Both

(d) None of these

10   Let $T : R^3 \rightarrow R^3$ be a linear transformation defined by $T(x,y,z) = (y,z,0)$, then the dimension of $R(T)$ is

(a) 0   (b) 2

(c) 1   (d) 3

11   If $T_1 : R^2 \rightarrow R^3$ and $T_2 : R^3 \rightarrow R^4$ are linear transformations, then $(T_2 \circ T_1)$ is linear transformation from

(a) $R^2 \rightarrow R^3$ (b) $R^3 \rightarrow R^2$

(c) $R^2 \rightarrow R^4$ (d) $R^4 \rightarrow R^2$

12   What is domain and co-domain for $(T_2 \circ T_1)(x,y)$, where $T_1(x,y) = (2x, 3y)$ and $T_2(x,y) = (x-y, x+y)$

(a) $(x-2y, y+2x)$ (b) $(2x-y, 3y+x)$

(c) $(2x-2y, 3x+3y)$ (d) $(2x-3y, 2x+3y)$

13   If $T(p,q,r) = (2p-q+3r, 4p+q-r)$, then $T(3,2,-1)$ is

(a) $(1,1)$ (b) $(1,15)$

(c) $(5,3)$ (d) $(15,1)$

14   If a function $T : R^3 \rightarrow R^3$ is defined on the basis $B = \{(1,1,1),(1,1,0),(1,0,0)\}$ of R^3 by $T(1,1,1) = (1,2,3)$, $T(1,1,0) = (2,3,4)$, $T(1,0,0) = (3,4,5)$, then $T(3,3,3)$ is

(a) $(1,2,3)$ (b) $(4,4,4)$

(c) $(3,6,9)$ (d) $(2,3,4)$

15   A linear transformation $T : R^m \rightarrow R^n$ is one-to-one if

(a) $\ker(T) = \{0\}$ (b) nullity $(T) \neq \{0\}$

(c) rank $(T) = n$ (d) none of these

16   An onto transformation is also called

(a) Injective transformation   (b) Bijective transformation

(c) Surjective transformation   (d) One-one

17   Let $T : R^4 \rightarrow R^4$ be an injective linear transformation, then T is

(a) Onto   (b) Not onto

(c) One-one   (d) Can't say

18   $T : R^m \rightarrow R^n$ is onto if

(a) rank $(T) = \dim(R^n)$ (b) rank $(T) = m$

(c) $\ker(T) = \{\ \}$ (d) $\ker(T) = \{0\}$

19   Standard matrix of a dilation with factor $k = 2$, followed by a rotation of $45°$, followed by a reflection about $y - axis$ is

(a) $\begin{bmatrix} \sqrt{2} & -\sqrt{2} \\ \sqrt{2} & \sqrt{2} \end{bmatrix}$ (b) $\begin{bmatrix} -\sqrt{2} & 5 \\ 0 & 1 \end{bmatrix}$

(c) $\begin{bmatrix} -\sqrt{2} & -\sqrt{2} \\ -\sqrt{2} & -\sqrt{2} \end{bmatrix}$ (d) $\begin{bmatrix} 5 & \sqrt{3} \\ 4 & 2 \end{bmatrix}$

20   Let P_2 be the vector space of all polynomials of degree at most two over R the set of all real
      numbers). Let a linear transformation $T : P_2 \to P_2$ be defined by
      $T(a+bx+cx^2) = (a+b)+(b-c)x+(b+c)x^2$. Consider the following statements
      (I)   The null space of T is $\{\alpha(-1+x+x^2): \alpha \in R\}$.
      (II)   The range space of T is spanned by the set $\{1+x^2, 1+x\}$.
      (III) $T(T(1+x)) = 1+x^2$.
      (IV)   If M is the matrix representation of T with respect to the standard basis $\{1, x, x^2\}$ of P_2,
             then the trace of the matrix M is 3.
      Which of the above statements are true?
      (a) I and II only   (b) I, III, and IV only
      (c) I, II, and IV only   (d) II and IV only

ANSWERS TO THE MULTIPLE-CHOICE QUESTIONS

1. a	11. c
2. b	12. d
3. a	13. b
4. d	14. c
5. b	15. a
6. d	16. c
7. b	17. c
8. c	18. b
9. c	19. a
10. b	20. c

Bibliography

1. Howard Anton and Chris Rorres (2009). *Elementary Linear Algebra – Application Version,* 9th edition, 390–442, Wiley, New Delhi, India.
2. Ravish R Singh and Mukul Bhatt (2014). *Linear Algebra and Vector Calculus,* 2nd edition, 3.1–3.86, McGraw-Hill, New York.
3. Kenneth Hoffman and Ray Kunze (1971). *Linear Algebra*, 2nd edition, 67–111, Prentice Hall Inc.
4. https://yutsumura.com/a-linear-transformation-is-injective-one-to-one-if-and-only-if-the-nullity-is-zero/

Index

For Product Safety Concerns and Information please contact our EU
representative GPSR@taylorandfrancis.com
Taylor & Francis Verlag GmbH, Kaufingerstraße 24, 80331 München, Germany

www.ingramcontent.com/pod-product-compliance
Lightning Source LLC
Chambersburg PA
CBHW061612220326
41598CB00024BC/3558